Schiff Base in Organic, Inorganic and Physical Chemistry

Edited by Takashiro Akitsu

Published in London, United Kingdom

Schiff Base in Organic, Inorganic and Physical Chemistry
http://dx.doi.org/10.5772/intechopen.104134
Edited by Takashiro Akitsu

Contributors
Ramhari Meena, Nighat Fahmi, Anita Kumari, Pooja Meena, Naveen Sharma, Dileep Ramakrishna, Shakeel Nawaz Shariff, Supriya Saravu, Mirjana M. Radanovic, Berta Barta Holló, Nuriye Tuna Subasi, Ruchi Shivhare, Kishor Danao, Deweshri Nandurkar, Vijayshri Rokde, Ashwini Ingole, Amol Warokar, Ujwala Mahajan, Adebimpe D. Adesina, Narjala Rama Jyothi, Snehalatha Pulivarthi, Shilpa Laxman Sangle, Sabeel M Basheer, Puthiyavalappil Rasin, Anandaram Sreekanth, Vipin Manakkadan, Vishnunarayanan Namboothiri Vadakkedathu Palakkeezhillam

First published in London, United Kingdom, 2023 by IntechOpen
IntechOpen is the global imprint of INTECHOPEN LIMITED, registered in England and Wales, registration number: 11086078, 5 Princes Gate Court, London, SW7 2QJ, United Kingdom

British Library Cataloguing-in-Publication Data
A catalogue record for this book is available from the British Library

Additional hard and PDF copies can be obtained from orders@intechopen.com

Schiff Base in Organic, Inorganic and Physical Chemistry
Edited by Takashiro Akitsu
p. cm.
Print ISBN 978-1-80355-678-9
Online ISBN 978-1-80355-679-6
eBook (PDF) ISBN 978-1-80355-680-2

For EU product safety concerns:
IN TECH d.o.o., Prolaz Marije Krucifikse Kozulić 3, 51000 Rijeka, Croatia,
info@intechopen.com or visit our website at intechopen.com.

Meet the editor

Takashiro Akitsu, Ph.D., is a professor in the Department of Chemistry, Faculty of Science Division II, Tokyo University of Science, Japan. He obtained his Ph.D. in Physical and Inorganic Chemistry from Osaka University, Japan, in 2000. Dr. Akitsu studied at the Institute for Protein Research, Osaka University (metalloproteins), Keio University, Japan (photo and magnetic functional organic/inorganic hybrid compounds), and Stanford University, USA (physical and bioinorganic chemistry) before moving to the Tokyo University of Science. He has published 220 articles and book chapters. He has also served as an editorial board member and peer reviewer for many journals and was involved in the organizing committees for several international conferences.

Contents

Preface IX

Chapter 1 1
Introduction to Schiff Base
by Shilpa Laxman Sangle

Chapter 2 15
Overview of Schiff Bases
by Nuriye Tuna Subasi

Chapter 3 37
Synthesis of Schiff Bases by Non-Conventional Methods
by Adebimpe D. Adesina

Chapter 4 47
Schiff Bases and Their Metal Complexes: Synthesis, Structural Characteristics and Applications
by Ramhari Meena, Pooja Meena, Anita Kumari, Naveen Sharma and Nighat Fahmi

Chapter 5 71
Some Aromatic Schiff Bases and Their Metal Complexes
by Mirjana M. Radanović and Berta Barta Holló

Chapter 6 103
Schiff Base as Multifaceted Bioactive Core
by Ruchi Shivhare, Kishor Danao, Deweshri Nandurkar, Vijayshri Rokde, Ashwini Ingole, Amol Warokar and Ujwala Mahajan

Chapter 7 131
Cytotoxic Activity of Schiff Bases and Their Complexes
by Narjala Rama Jyothi and Snehalatha Pulivarthi

Chapter 8 **153**
Recent Advancements in Schiff Base Fluorescence Chemosensors for the Detection of Heavy Metal Ions
by Sabeel M. Basheer, Puthiyavalappil Rasin, Vipin Manakkadan, Vishnunarayanan Namboothiri Vadakkedathu Palakkeezhillam and Anandaram Sreekanth

Chapter 9 **175**
Schiff Base Complexes for Catalytic Application
by Shakeel Nawaz Shariff, Supriya Saravu and Dileep Ramakrishna

Preface

The Schiff base is a well-known organic compound category and a typical organic ligand for metal complexes in both organic and inorganic chemistry. However, development in material and life sciences using Schiff base compounds is still ongoing. This book provides a comprehensive overview of Schiff base chemistry.

Each chapter presents various viewpoints of Schiff base in current developments: overview and synthesis of organic chemistry; metal complexes with structures, aromatic ligand, bioactive core, cytotoxic activity as inorganic chemistry, chemosensors for heavy metal ions and catalytic applications as applied chemistry. The information summarized in this book will be useful for all readers who are interested in Schiff base and their metal complexes.

From where is such a variety? Indeed, imine or azomethine (C=N) of Schiff base is merely a functional group, which is usually synthesized from aldehyde and (primary) amine. However, introducing extended groups into aldehyde or amine is easy. Therefore, as a lone pair donating group (Lewis base) coordinating with a metal ion to yield metal complexes, Schiff base (ligands) and their metal complexes can obtain different functions or properties. For example, azobenzene containing salen-type metal complexes have been used as photo-functional dyes dispersing into polymer films. Besides coordination to metal ions as a rigid chelate ligand, introducing an azobenzene group and strong pi-conjugated planar moiety plays an essential role in polarized light-induced control of the molecular orientation, which was also investigated with DFT calculation theoretically. As for material science use, the stability of Schiff base ligands has advantages. We used redox active Schiff base metal complexes (potentially exhibiting catalytic function) for electron transfer mediator to a metalloenzyme. In this case, the reactivity of the Schiff base metal complex plays an important role. As for catalytic or biochemical reactions towards assisting life science, reactivity should be considered during molecular design.

At glance, stability and reactivity are opposite requirements, though Schiff base (ligands) can realize both by molecular design. Thus, Schiff base compounds have wide applications and are at the frontier of chemistry.

Takashiro Akitsu
Department of Chemistry,
Tokyo University of Science,
Tokyo, Japan

Chapter 1

Introduction to Schiff Base

Shilpa Laxman Sangle

Abstract

When any primary amine reacts with an aldehyde or a ketone under specific conditions, Schiff bases are formed. The general structure of Schiff base is R2C=NR′ and is considered as a subclass of imines which is commonly used as a synonym for azomethine. The first imines were prepared in the nineteenth century by a classical method that involves condensation of a carbonyl compound with the help of amine under the distillation of azeotropic and to remove water formed in the system, molecular sieves are used. Later many ways of synthesis of Schiff bases are invented. Schiff bases exhibit a wide range of biological activities and are commonly used for industrial purposes. These are the most widely used as intermediates in organic synthesis, catalysts, pigments and dyes, polymer stabilizers, etc.

Keywords: Schiff bases, biological activity of Schiff bases, azomethine, Schiff base metal complexes, formation mechanism of Schiff base

1. Introduction

In the year 1864, Hugo Schiff was the first to synthesize Schiff's base under azeotropic distillation by using aldehyde or ketone and primary amine. They can be considered a sub-class of imines with the general structure $R_1R_2C{=}NR'$ ($R' \neq H$) [1–5]. Depending on their structure, they can be considered as either secondary aldimines or secondary ketimines. When these compounds are being used as ligands to form coordination complexes with metal ions, the term Schiff base is applied. Corrin complexes occur naturally, but the majority of artificial Schiff bases are used to form many important catalysts, such as Jacobsen's catalyst (**Figure 1**).

Schiff bases are imines in which R_3 is an alkyl or aryl group (not hydrogen). R_1 and R_2 may be hydrogen. Schiff bases have a wide range of biological properties such as antimicrobial, anticancer, and antiviral. Inhibition of amyloid-β aggregation is achieved by Schiff bases [6]. They are common enzymatic intermediates where an aldehyde or ketone of a cofactor or substrate reversibly reacts with the terminal group of a lysine residue. Lysine residue forms a Schiff base with the common enzyme cofactor pyridoxal phosphate (PLP) and is transaldiminated to the substrate(s) [7]. Similarly, the cofactor retinal forms a Schiff base in human rhodopsin (via Lysine 296), which is key in the photoreception mechanism.

In coordination chemistry, Schiff bases are common ligands. The ligands are derived from aromatic aldehydes and alkyl diamines [8]. The imine nitrogen is

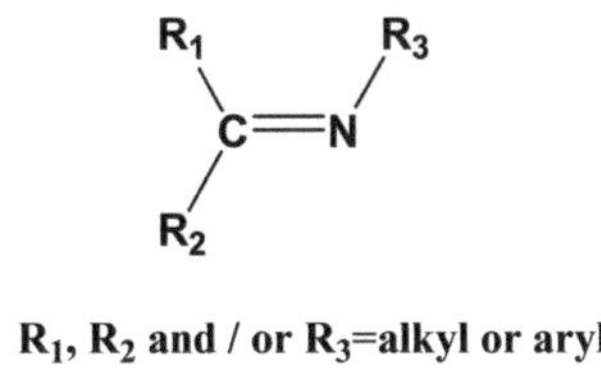

Figure 1.
General structure of an imine.

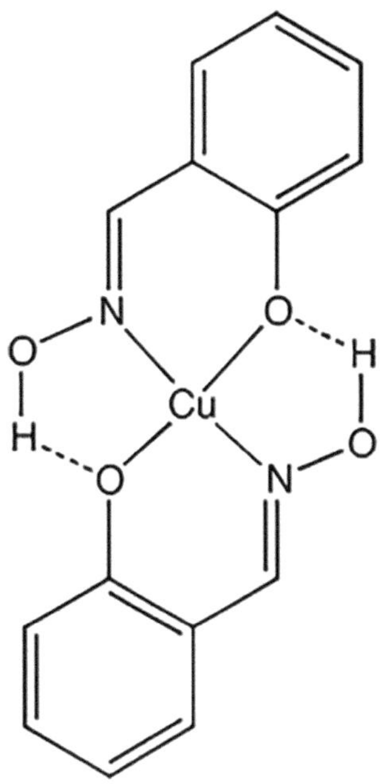

Figure 2.
Copper (II) complex of the Schiff base ligand salicylaldoxime.

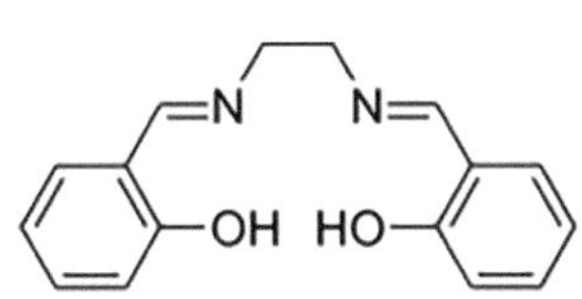

Figure 3.
Salen is a common tetradentate ligand that becomes deprotonated upon complexation.

basic in nature and exhibits pi-acceptor properties. In 1968, Ryōji Noyori was awarded a share of the 2001 Nobel Prize in Chemistry for the development of a copper-Schiff base complex for the metal-carbenoid cyclopropanation of styrene (**Figures 2–4**) [9].

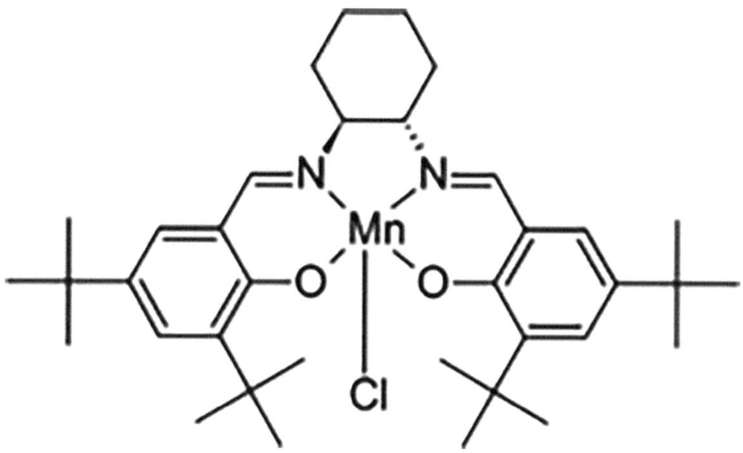

Figure 4.
Jacobsen's catalyst is derived from a chiral Salen ligand.

2. Schiff base ligands

Researchers' interest in Schiff bases prepared from ortho-hydroxyl aromatic aldehydes is due to their ability to act as bidentate ligands for transitional metal ions [10–14]. Later, in studies, it has been observed that azomethines from salicylaldehydes gave the best quantitative structure-antitumor activity relationship which has been studied for a series of Schiff bases derived from a variety of substituted aromatic amines and aldehydes [15, 16]. Schiff bases prepared from salicylaldehydes have also been reported as antimycotic or antimicrobial activity and plant growth regulators [17–19]. They also show some analytical applications [20]. The −N=CH− (imine) group is the characteristic of Schiff bases. Imine group imports in elucidating the mechanism of racemization and transamination reaction in biological systems [21, 22]. Schiff bases are active against many organisms such as Erysiphe graminis, Bacillus polymxa, Staphylococcus aureus, Candida *albicans*, Escherichia coli, Trychophyton gypseum, Plasmopora viticola, and Mycobacteria. They have shown excellent stability, and selectivity for specific metal ions such as Pb (II), Co (II), Al (III), Ag (II), Gd (III), Cu (II), Ni (II), Y(III), Zn (II), and Hg (II) [23–28], so that large number of different Schiff base ligands have been used in potentiometric sensors as cation carriers.

Due to the important properties of Schiff bases in catalysis [29], they have been studied. In the hydrogenation of olefins, schiff bases show catalytic activity [30] and applications in biomimetic catalytic reactions. An interesting application of Schiff bases based on their ability to spontaneously form a monolayer on the surface to be protected, as an effective corrosion inhibitor Schiff bases function more efficiently than many commercial inhibitors including aldehydes or amines due to the C=N bond [31]. The principal interaction between the metal surface and inhibitor [32] is Chemisorption. The inhibitor molecule should have centers that can form bonds with the metal surface by electron transfer. In such cases, the inhibitor acts as a Lewis base and the metal acts as an electrophile. The protective compound has oxygen and nitrogen atoms with free electron pairs which are readily available for sharing and serves as a nucleophilic center. They create multiple absorption sites for the inhibitor along with the atoms of the benzene rings thus enabling stable monolayer formation [33].

Imines also have biological importance. In the chemistry of vision, an imine linkage between the protein opsin in the retina of the eye and the aldehyde derived from vitamins plays an important role. Vitamins are large proteins that catalyze chemical changes in cells. They are also called coenzymes as vitamins do the functioning of many enzymes. Pyridoxal phosphate is a biologically important aldehyde that is the active form of vitamin B6. It serves as a coenzyme by forming an imine with an amino acid grouping an enzyme. In a transamination reaction, the coenzyme, bound to the enzyme, is involved. The transfer of the amino group from one amino acid to another, is important in biosynthesis of amino acids and the metabolism. In the enzyme-catalyzed hydrolysis, which is the last step, involves the cleavage of the imine to the modified amino acid and pyridoxal. Schiff bases biological properties, such as antifungal, antibacterial activities are reported [34–37]. Because of the anticancer and herbicidal applications [38, 39], their metal complexes have been widely studied. For biologically important species, they serve as models. For Schiff bases, we chose hydrazides, dihydrazides, hydrazones, and mixed derivatives such as hydrazide-hydrazones.

3. Synthesis of Schiff bases

The first imines were synthesized in the nineteenth century by Hugo Schiff in 1864. Since then, a variety of methods have been developed for the synthesis of imines [40]. The classical synthesis involves the condensation of a carbonyl compound with an amine under azeotropic distillation reported by Schiff. Water formed in the system was completely removed by using **a** molecular sieve [41]. An *in-situ* method for water elimination was developed using dehydrating solvents such as tetramethyl orthosilicate or trimethyl orthoformate in the 1990s [41, 42]. Chakraborti et al. [43] demonstrated that the efficiency of these methods is dependent on the use of highly electrophilic carbonyl compounds and strongly nucleophilic amines. Schiff bases are often polydentate in coordinating ability, because of synthetic flexibility, the special property of C=N group and the relative ease of preparation, especially when −SH or −OH are present close to the azomethine group which can form a five or six membered ring with the metal ion (**Figures** 5 and **6**).

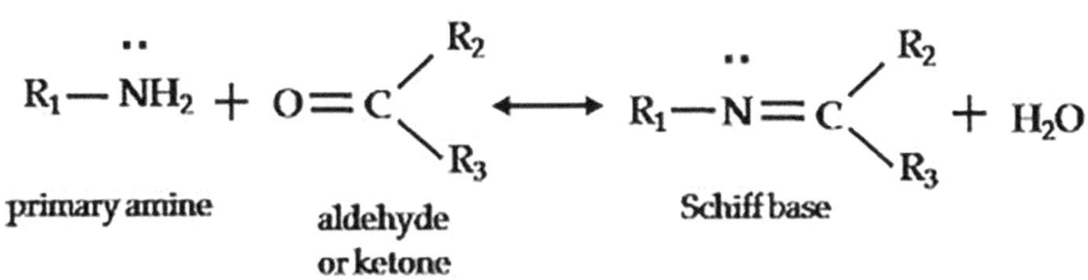

Figure 5.
General formation of Schiff base.

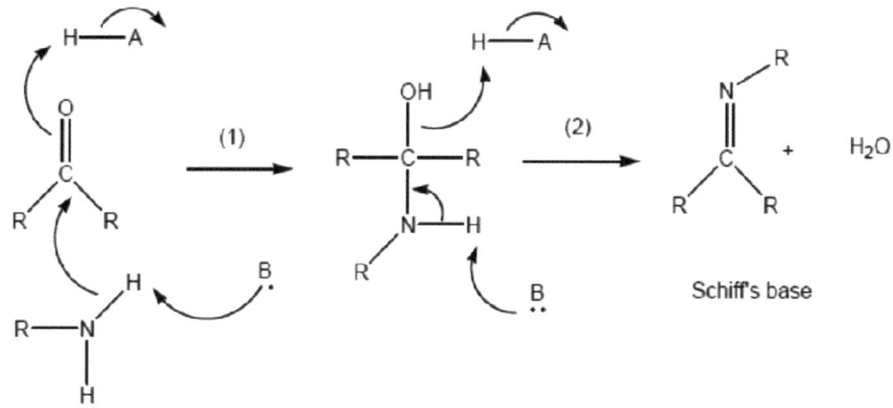

Figure 6.
General mechanism.

3.1 Denticity and basicity of Schiff base

The Schiff base ligands are classified according to the number of donor atoms and are named as uni-, di-, tri-, and tetra-dentate ligands. Schiff bases possess nitrogen donor atoms, so can act as bi-, tri-, tetra- or polydentate ligand. In general the donor nature of the ligand depends both on the type of aldehyde/ketone used and the nature of primary amine/diamine.

The basicity of the Schiff base also plays a key role in the formation and stabilization of the complexes. The −OH group present in the Schiff base can induce tautomerism in the compound, which leads to a compound with different structures. A large number of Schiff base compound show keto-enol tautomerism. Also, the deprotonation of alcoholic and phenolic groups is favored due to the stabilization of various oxidation states of different metal ions. Coordination with transition metal Schiff base metal complexes is prepared *in situ* by producing a reaction between the Schiff base and well-defined metal. This approach is clearly simple and suitable for catalytic applications. Different concentrations of different complexes can be present, when an equilibrium constant is expressed as a concentration quotient. However, the identity and homogeneity of the complex can be controlled by the introduction of a bulky group in the Schiff bases due to the shifting of the equilibrium toward the formation of a single species. A disproportionation between Schiff base metal complexes and the metal alkoxides can occur and the stability of the complexes is regulated by the equilibrium constant.

Schiff base ligands are able to coordinate many different metals with various oxidation states, enabling the use of Schiff base metal complexes for a large variety of useful catalytic transformations. Schiff-base ligands containing imidazole groups have potential donor and acceptor character in the formation of a coordination bond and function as a ligand-complex or as a self-complementary building block for the construction of the assembly structure due to the formation of a coordination bond with Cu (II) ions [43, 44]. The versatility of Schiff base ligands and the biological, analytical, and industrial applications of their complexes have promoted further investigations in this area. The importance of Schiff base complexes for bioinorganic chemistry, biomedical applications, supramolecular chemistry, catalysis and material science, separation and encapsulation processes, and formation of compounds with unusual properties and structures has been well recognized and reviewed. A large number of Schiff bases and their complexes are of significant attention because of their biological activity including antitumor, antibacterial, fungicidal, and anticarcinogenic properties and catalytic activity [45, 46].

4. Application of Schiff base complexes

4.1 Catalysis

The Schiff base transition metal complexes are cheap, easy to synthesize, and their chemical and thermal stability [47] make them a family of attractive oxidation catalysts for a variety of organic substrates. Important oxidation reactions include the oxidation of sulfides to sulfoxides, the activation of hydrocarbons, alkenes to epoxides and diols, and the transformation of alcohols to either the corresponding carbonyl compounds or carboxylic acids. The catalytic activities of the Cu (II), Co (II), Fe (III), and Mn (II) complexes are observed for the phenol hydroxylation reaction. The activities of these cobalt complexes are slightly lower than that of manganese (II), iron (II), and copper (II) analogs of the investigated Schiff bases [48, 49]. Catechol was found as the major product of the reaction [50]. Due to the dimer formation, the cobalt (II) complex is found to be inactive. So, it is unable to form the intermediate by binding with oxygen. The most active catalyst found was the copper complex.

The environment at the coordination center in Schiff base metal complexes can be modified by attaching different substituents to the ligand and a useful range of electronic and steric properties essential for the fine-tuning of reactivity and structure can be achieved [51–53]. Metal complexes of Schiff bases with p-block and d-block metals have been known to act as highly efficient catalysts in various reactions and other useful syntheses [54–60].

In the syntheses of quality polymers, many Schiff base complexes of ruthenium and palladium are used as a catalyst. Katsuki reviewed the unique asymmetric catalysis of metal complexes of Salen and the related Schiff-base ligands [51]. The review summarizes the formation of cis metallo-salen and its related complexes, their structural features, and their application to asymmetric syntheses. In 1999, the effective oxidation of olefins using Mn (II) amino acid Schiff base complexes was reported by Wang et al. [58]. The catalytic activities of transition metal complexes-both simple and polymer anchored were reviewed by Gupta and Sutar. They have highlighted the use of Schiff base complex as a catalyst for ring closures, hydrogenations, various coupling reactions, oxidations, and polymerizations [61, 62]. Due to the better selectivity and recyclability of homogeneous and heterogenous catalysts, they have recently attracted the attention of chemists. In recent years, the number of publications in catalysis supported by Schiff base complexes was exponentially increased. However, homogeneous catalysis is more relevant due to the mechanism of the reaction that can be arrived at. Ligand famous for their stereoselective transformations is BINAP ligands (BINAP is the abbreviation for the organophosphorus compound 2,2′- bis (diphenylphosphino)-1,1′binaphthyl). The catalytic activity of chiral BINAP Schiff base complexes in stereoselective organic transformations has been reviewed by Cheand Huang [63]. Their studies reveal that these types of chiral metal complexes are active catalysts for stereoselective organic transformations including aldol reactions, ring-opening polymerization of lactide, hydroxylation of styrene, Diels-Alder reactions of 1,2-dihydropyridine, trimethylsilyl cyanation of aldehydes, Baeyer-Villiger oxidation of aryl cyclobutanone, desymmetrization of meso-N-sulfonylaziridine, and alkene epoxidation.

4.2 Biological activity

Schiff bases exhibit a broad range of biological activities, including antiviral, antipyretic properties, antimalarial, antibacterial, anti-inflammatory, antifungal, and antiproliferative [64]. In various natural, natural-derived, and non-natural compounds, imine or azomethine groups are present. The imine group present is critical to their biological activities in such compounds [65]. Schiff bases are interesting moieties for the design of antimalarial agents. Schiff base was the most effective antimalarial agent among the synthesized 5-nitroisoquinoline derivatives. Ancistrocladidine is a secondary metabolite produced by plants from the families of Ancistrocladaceae and Dioncophyllaceae that present an imine group in its molecular scaffold. Rathelot et al. [66] described the synthesis of Schiff-base functionalized 5-nitroisoquinolines and investigated the *in vitro* activity of these compounds. Isatin-derived Schiff bases have also been reported to possess antibacterial activity. Other isatin derivative Schiff bases also have antibacterial activities [67, 68]. The natural or non-natural origins that are platforms for the synthesis of Schiff bases for antibacterial activities include amino acids, coumarins, sulfonamides, or resacetophenone, aminothiazolyl bromocoumarin, crown ethers, o-phthaldehyde, or 2-aminophenol and 1,2,4-triazoles [69]. Kumar et al. [70] reported a series of 3-(benzylidene amino)-2-phenylquinazoline-4(3H)-one and evaluated their cytotoxicity and antiviral activity. Compounds having 2-hydroxy substitution showed better antiviral activity. Some bis-Schiff bases of isatin, benzylisatin, and 5-fluoroisatin, were reported by Jarrahpour et al. [68] as antiviral agents. Sashidhara et al. [71] presented a series of Schiff's bases of benzocoumarin and evaluated *in vitro* for their antioxidant activity and *in vivo* for their antidyslipidemic activity. During antioxidant screening, the compound exhibited significant activity and significant lipid-lowering activity. Ferrocenyl Schiff bases antioxidant capacities including o-(1 ferrocenyl ethylidene amino) phenol (OFP), m-(1-ferrocenyl ethylidene amino) phenol (MFP), and p-(1-ferrocenyl ethyl idene amino) phenol (PFP) were evaluated OFP. MFP and PFP possessed similar activities to trap DPPH and $ABTS^{+}$, which was reported by Li et al. [72]. The antioxidant effectiveness of Schiff base increased by the introduction of the ferrocenyl group more remarkably than benzene-related Schiff bases.

4.2.1 Ribonucleotide reductase

While forming chelates with essential metal ions, thiosemicarbazones in their neutral or deprotonated form, act as N, N, S-thiodentate ligands. On different tumor cell lines, they display antiproliferative activity. A strong correlation has been established between the enzyme Ribonucleotide Reductase (RR), a necessary enzyme for DNA synthesis, and tumor growth rate [73]. In the 60s, the first antitumor effect of thiosemicarbazones was obtained and deserves a brief resume. In 1956, Brockman et al. first reported the antileukemic effect of 2-formyl pyridine thiosemicarbazone [61]. The hypotheses about the mode of action of the α-(N)-heterocyclic thiosemicarbazones, the active molecules with tridentate nature, that allows them to be effective chelators, and a better activity was obtained by modifying the aromatic system was formulated in 1965, by French et al. [62]. The activity of 1-formyl isoquinoline thiosemicarbazone and pyrazine carboxaldehyde thiosemicarbazone was predicted based on this principle. The reduction of ribose to deoxyribose through a free radical mechanism that is triggered by a tyrosyl radical was promoted by Ribonucleotide reductase, which is an iron-dependent enzyme. The synthesis phase of the cell cycle was blocked due to the inhibition of this enzyme and eventually cell death by apoptosis. The active species was the iron (II) complex of 1-formyl isoquinoline thiosemicarbazone, which is also indirectly demonstrated by them. In fact, it was later discovered that iron and copper complexes are more active than the free ligands [63]. Thelander et al. proposed a reasonable mechanism [74] which proved, by exposing ribonucleotide reductase to the aforementioned molecules, that the drug targeted the tyrosyl free radical of the enzyme which was inhibited by the thiosemicarbazone complex by destroying the radical. This mechanism excludes the role of thiosemicarbazones as

Structure	Biological application	Reference
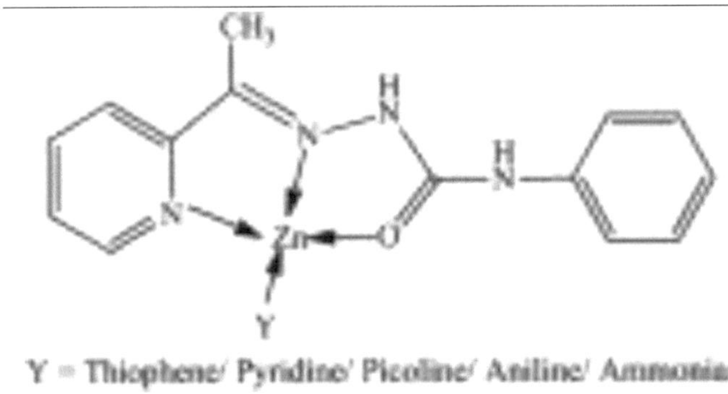	Antimicrobial activity against Staphylococcus aureus, Bacillus anthracis, Aspergillus niger, and Candida albicans	Int. J. Scientific Technology Res., 3 (2014) 73–77.
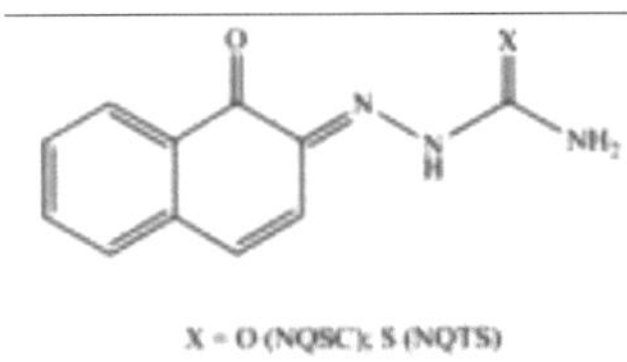	Anticancer activity against MCF-7 breast cancer cell lines	J. Inorg. Biochem., 99 (2005) 1526–1531.
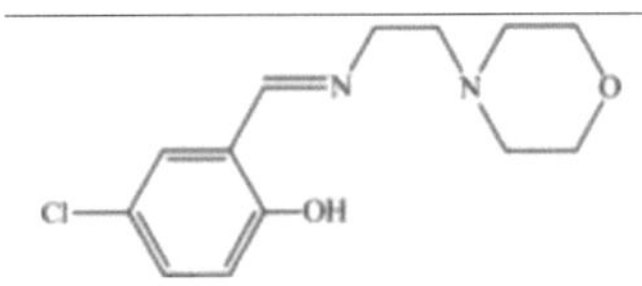	Antibacterial activity against Gram +ve bacterial (Bacillus subtilis and S. aureus) and Gram –ve bacterial (E. coli and P. fluorescence)	J. Coord. Chem., 62 (2009) 3471–3477.
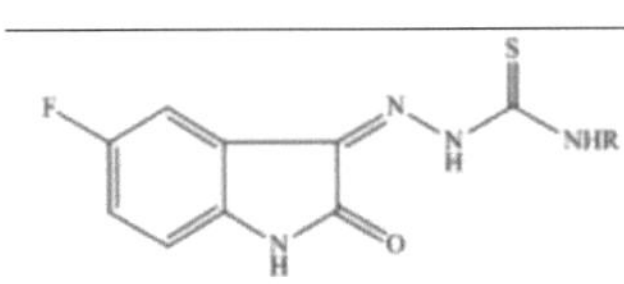	Antiviral activity, Inhibits the replication of vesicular stomatitis virus and shows cytotoxicity in Vero clone CCL-81 cell lines	Monatsh Chem.,144 (2013) 1725

Table 1.
Various semi-carbazones, Schiff bases, and their biological applications.

simple iron chelators and requires oxygen. This is in agreement with the experimental observations and the reaction is reversible, as reported by them. In the enzyme, there must be a hydrophobic pocket or patch with which the aromatic system interacts, which indicates that methylation on the aromatic ring of 2-formyl pyridine thiosemicarbazone renders this compound more active. It proves that 1-formyl isoquinoline thiosemicarbazone inhibits more strongly ribonucleotide reductase than 2-formyl pyridine thiosemicarbazone. It was identified by Agrawal et al. [75] while searching for an optimum bulk for the aromatic fragment, the most active compound found in the quinoline series was 2-formyl-4-(3-amino) phenyl pyridine thiosemicarbazone instead 1-formyl-5-aminoisoquinoline thiosemicarbazone (**Table 1**) [75–77].

5. Conclusion

Schiff bases have been widely used in various organic transformations and for industrial applications. However, further investigation is required about the biological activity of this class of compound. This becomes clear when plant pathogens are considered. Recently, there is an increase in the number of reports including the effects of the Schiff bases on the pathogens of clinical interest. The promising leads for the design of more efficient antimicrobial agents have been shown by Schiff base compounds. Analyses of the mechanism of action of these compounds as well as the structure-activity relationships of the Schiff bases require advanced research in this field.

Author details

Shilpa Laxman Sangle
Department of Chemistry, MVP Samaj's SSSM Arts, Science and Commerce College (Affiliated to S.P. Pune University), Saikheda, India

*Address all correspondence to: shilpasangle620@gmail.com

References

[1] IUPAC, editor. Compendium of Chemical Terminology. 2nd ed. IUPAC; Wiley–Blackwell. 11 August 1997

[2] Schiff H. Mittheilungen aus dem Universitäts-laboratorium in Pisa: 2. Eine neue Reihe organischer Basen [Communications from the university laboratory in Pisa: 2. A new series of organic bases]. Annalen der Chemie und Pharmacie (in German). 1864;**131**:118-119. DOI: 10.1002/jlac.18641310113

[3] Schiff U. Sopra una nova serie di basi organiche [On a new series of organic bases]. Giornale di Scienze Naturali ed Economiche (in Italian). 1866;**2**:201-257

[4] Schiff H. Eine neue Reihe organischer Diamine [A new series of organic diamines]. Annalen der Chemie und Pharmacie. Supplementband (in German). 1866;**3**:343-370

[5] Schiff H. Eine neue Reihe organischer Diamine. Zweite Abtheilung [A new series of organic diamines. Second part.]. Annalen der Chemie und Pharmacie (in German). 1866;**140**:92-137. DOI: 10.1002/jlac.18661400106

[6] Bajema EA, Roberts KF, Meade TJ. Chapter 11. Cobalt-Schiff Base complexes: Preclinical research and potential therapeutic uses. In: Sigel A, Freisinger E, Sigel RKO, Carver PL, editors. Essential Metals in Medicine: Therapeutic Use and Toxicity of Metal Ions in the Clinic. Berlin: de Gruyter GmbH; 2019. pp. 267-301

[7] Eliot AC, Kirsch JF. Pyridoxal phosphateenzymes: Mechanistic, structural, and evolutionary considerations. Annual Review of Biochemistry. 2004;**73**:383-415

[8] Hernández-Molina R, Mederos A. Acyclic and macrocyclic Schiff Base ligands. Comprehensive Coordination Chemistry. 2003;**II**:411-446

[9] Nozaki H, Takaya H, Moriuti S, Noyori R. Homogeneous catalysis in the decomposition of diazo compounds by copper chelates: Asymmetric carbenoid reactions. Tetrahedron. 1968;**24**(9):3655-3669. DOI: 10.1016/S0040-4020(01)91998-2

[10] Pfeiffer P, Breith E, Llibbe E, Tsumaki T. Justus Liebigs Annalen. Tricyclische orthokondensierte Nebenvalenzringe der Chemie. 1933;**503**:84-130. DOI: 10.1002/jlac.19335030106

[11] Hunter L, Marriott JA. Co-ordinated copper and nickel compounds of salicylidene derivatives. Journal of the Chemical Society. 1937;**1937**:2000

[12] Sacconi L, Ciampolini M, Maggio F, Cavasini FP. Studies in Coordination Chemistry. IX.1 Investigation of the Stereochemistry of Some Complex Compounds of Cobalt (II) with N-Substituted Salicylaldimines. Journal of the American Chemical Society. 1962;**84**:3246-3248

[13] Holm RH, Swaminathan K. Studies on Nickel (II) Complexes. III. Bis-(N-arylsalicylaldimine) Complexes. Inorganic Chemistry. 1962;**1**:599-607

[14] Perry GC, Thornton DA. N-aryl salicylaldimine complexes: Infrared and PMR spectra of the ligands and vibrational frequencies of their metal (II) chelates. Journal of Inorganic Chemistry. 1972;**34**:3357-3367

[15] Hodnett EM, Dunn WJ. Journal of Medicinal Chemistry. 1970;**13**:768

[16] Hodnett EM, Mooney PD. Journal of Medicinal Chemistry. 1970;**13**:786

[17] G.H. Alt (Monsanto Co.), US. 4.226.615 (1980)

[18] Hamada Y, Takeuchi I, Ita Y, Matsui S, Ita T. Yakugaku Zasslzi. 1981;**101**:633

[19] Ismail M. Indian Journal of Pharmaceutical Science. 1986;**45**:121

[20] Palet PR, Thaker BT, Zele S. Indian Journal of Chemistry. 1999;**A38**:563

[21] Lau KY, Mayr A, Cheung KK. Inorganic Chemistry Acta. 1999;**285**:223

[22] Shawali AS, Harb NMS, Badahdah KO. Journal of Heterocylic Chemistry. 1985;**22**:1397

[23] Abbaspour A, Esmaeilbeig AR, Jarrahpour AA, Khajeh B, Kia R. Talanta. 2002;**58**:397

[24] Mahajan RK, Kaur I, Kumar M. Sensors and Actuators, B-91. 2003;**26**

[25] Ganjali MR, Golmohammadi M, Yousefi M, Norouzi P, Salavati-Niasari M, Javanbakht M. Analytical Science. 2003;**19**:223

[26] Jain AK, Gupta VK, Ganeshpure PA, Raisoni JR. Analytica Chimica Acta. 2005;**553**:177

[27] Jeong T, Lee HK, Jeong DC, Jeon S. Talanta. 2005;**65**:543

[28] Gupta VK, Singh AK, Mehtab S, Gupta B. Analytica Chimica Acta. 2006;**566**:5

[29] Hernandes MM, Mckee ML, Keizer TS, Yeaswood BC, Atwood DA. Journal of Chemical Society. 2002;**410**

[30] Olie GH, Olive S. Springer, Berlin 1984

[31] Li S, Chen S, Ma H, Yu R, Liu D. Corrosion Science. 1999;**41**:1273

[32] Ashassi-Sorkhabi H, Shabani B, Aligholipour B, Seifzadeh D. Applied Surface Science. 2006;**252**:4039

[33] Quan Z, Chen S, Li Y. Corrosion Science. 2001;**43**:1071

[34] Williams DR. Chemical Reviews. 1972;**72**:203

[35] Campos A, Anacona JR, Campos-Vallette MM. Mian Group Metal Chemistry. 1999;**22**:283

[36] Sari N, Arslan S, Logoglu E, Sakiyan I. Science. 2003;**16**:283

[37] Verma M, Pandeya SN, Singh KN, Stabler JP. Acta Pharm. 2004;**54**:49

[38] Cozzi PG. Chemical Society Reviews. 2004;**410**

[39] Chandra S, Sangeetika J. Journal of the Indian Chemical Society. 2004;**81**:203

[40] Zheng Y, Ma K, Li H, Li J, He J, Sun X, et al. Catalysis Letters. 2009;**128**:465

[41] Taguchi K, Westheimer FH. The Journal of Organic Chemistry. 1971;**36**:1570

[42] Love BE, Ren J. The Journal of Organic Chemistry. 1993;**58**:5556

[43] Chakraborti AK, Bhagat S, Rudrawar S. Tetrahedron Letters. 2004;**45**:7641

[44] Cozzi PG. Chemical Society Reviews. 2004;**33**:410

[45] Sunatsuki Y, Motoda Y, Matsumoto N. Coordination Chemistry Reviews. 2002;**226**:199

[46] Ren S, Wang R, Komatsu K, Bonaz-Krause P, Zyrianov Y,

Mckenna CE, et al. Journal of Medical Chemistry. 2002;**45**:410

[47] Sheldon RA, Kochi JK. Metal-Catalyzed Oxidations of Organic Compounds. New York: Academic Press; 1981

[48] Van Wyka JL, Mapolie SF, Lennartson A, Håkansson M, Jagner S. Inorganic Chimica Acta. 2008;**361**:2094

[49] Van Wyk JJL, Mapolie SF, Lennartson A, Håkansson M, Jagner S. Z. Naturforch. 2007;**62**:331

[50] Park JN, Wang JK, Choi Y, Dong WY, Hong SJ, Lee CW. Journal of Molecular Catalysis A. 2006;**247**:73

[51] Che CM, Huang JS. Metal complexes of chiral binaphthyl Schiff-base ligands and their application in ~ 2866 ~ international journal of chemical studies stereoselective organic transformations. Coordination Chemistry Reviews. 2003;**242**:97

[52] Canali L, Sherrington DC. Utilisation of homogeneous and supported chiral metal (salen) complexes in asymmetric catalysis. Chemical Society Reviews. 1999;**28**:85

[53] Gibson VC, Spitzmesser SK. Advances in non-metallocene olefin polymerization catalysis. Chemical Reviews. 2003;**103**(1):283

[54] Rao SN, Kathale N, Rao NN, Munshi KN. Catalytic air oxidation of olefins using molybdenum dioxo complexes with dissymmetric tridentate O, N, S-donor Schiff base ligands derived from o-hydroxyacetophenone and S-benzyldithiocarbazate or S-methyldithiocarbazate. Inorganica Chimica Acta. 2007;**360**(14):4010

[55] Jammi S, Saha P, Sanyashi S, Sakthivel S, Punniyamurthy T. Chiral binuclear copper (II) catalyzed nitroaldol reaction: Scope and mechanism. Tetrahedron. 2008;**64**(51):11724

[56] Iwakura I, Ikeno T, Yamada T. Proposal for the metallacycle pathway during the Cyclopropanation catalyzed by cobalt–Schiff Base complexes. Organic Letters. 2004;**6**(6):949

[57] Katsuki T. Unique asymmetric catalysis of cis-β metal complexes of salen and its related Schiff-base ligands. Chemical Society Reviews. 2004;**33**(7):437

[58] Wang RM, Hao CJ, Wang YP, Li SB. Amino acid Schiff base complex catalyst for effective oxidation of olefins with molecular oxygen. Journal of Molecular Catalysis A: Chemical. 1999;**147**:173

[59] Gupta KC, Sutar AK. Catalytic activities of Schiff base transition metal complexes. Coordination Chemistry Reviews. 2008;**252**(12-14):1420

[60] Gupta KC, Sutar AK, Lin CC. Polymer-supported Schiff base complexes in oxidation reactions. Coordination Chemistry Reviews. 2009;**253**(13-14):1926

[61] Brockman RW, Thomson JR, Bell MJ, Skipper HE. Observations on the antileukemic activity of pyridine-2-carboxaldehyde thiosemicarbazone and thiocarbohydrazone. Cancer Research. 1956;**16**:167-170

[62] French FA, Blanz EJJ. The carcinostatic activity of alpha-(N) heterocyclic carboxaldehyde thiosemicarbazones. I. Isoquinoline-1-carboxaldehyde thiosemicarbazone. Cancer Research. 1965;**25**:1454-1458

[63] Antholine WE, Knight JM, Petering DH. Inhibition of tumor cell trans plantability by iron and copper

complexes of 5-substituted 2- formyl pyridine thiosemicarbazones. Journal of Medicinal Chemistry. 1976;**19**:339-341

[64] Anand P, Patil VM, Sharma VK, Khosa RL, Masand N. International Journal of Drug Design and Discovery. 2012;**3**:851

[65] Da Silva CM, Da Silva DL, Modolo LV, Alves RB, de Resende MA, Martins CV, et al. Journal of Advanced Research. 2011;**2**:1

[66] Rathelot P, Vanelle P, Gasquet M, Delmas F, Crozet MP, Timon-David P, et al. European Journal of Medical Chemistry. 1995;**30**:503

[67] Pandeya SN, Sriram D, Nath G, de Clercq E. European Journal of Pharmaceutical Sciences. 1999;**9**(1):2

[68] Jarrahpour A, Khalili D, De Clercq E, Salmi C, Brunel JM. Molecules. 2007;**12**(8):1720

[69] Bayrak H, Demirbas A, Karaoglu SA, Demirbas N. European Journal of Medicinal Chemistry. 2009;44(3):1057

[70] Kumar KS, Ganguly S, Veerasamy R, Clercq ED. European Journal of Medicinal Chemistry. 2010;**45**:5474

[71] Sashidhara KV, Rosaiah JN, Bhatia G, Saxena JK. Synthesis, anticancer activities and experimental-theoretical DNA interaction studies of 2-amino-4-phenyl-4H-benzo[h] chromene-3-carbonitrile, European Journal of Medicinal Chemistry. 2008;**43**(2592):37

[72] Li YF, Zai-Qun L. Ferrocenyl Schiff base as novel antioxidant to protect DNA against the oxidation damage. European Journal of Pharmaceutical Sciences. 2011;**44**:158

[73] Khan T, Ahmad R, Joshi S. Anticancer potential of metal thiosemicarbazone complexes: A review. Der Chemica Sinica. 2015;**6**(12):1-11

[74] Thelander L, Gräslund A. Mechanism of inhibition of mammalian ribonucleotide reductase by the iron chelate of 1-formylisoquinoline thiosemicarbazone. Destruction of the tyrosine free radical of the enzyme in an oxygen-requiring reaction. Journal of Biological Chemistry. 1983;**258**:4063-4066

[75] Agrawal KC, Mooney PD, Sartorelli AC. Potential antitumor agents. 4-Methyl-5-amino-1-formylisoquinoline thiosemicarbazone. Journal of Medicinal Chemistry. 1976;**19**:970-972

[76] Agrawal KC, Lin AJ, Booth BA, Wheaton JR, Sartorelli AC. Potential antitumor agents. 9. 2-Formyl(m-amino) phenyl pyridine thiosemicarbazones. Journal of Medicinal Chemistry. 1974;**17**:631-635

[77] Pelosi G. Thiosemicarbazone metal complexes: From structure to activity. The Open Crystallography Journal. 2010;**3**:16-28

Chapter 2

Overview of Schiff Bases

Nuriye Tuna Subasi

Abstract

Schiff bases, which were first obtained by the German chemist H. Schiff in 1864, are used in the paint industry, polymer technology, pharmaceutical industry, medicine, agriculture, preparation of rocket fuel, and explanation of biological events, and in many other areas due to the groups in their structures. This chapter will be a guide that contains a summary of general information that should be known about these compounds, which have a wide range of use in our daily life. In this chapter, the following topics are planned to be explained. (1) Schiff bases, physical and chemical properties, (2) the formation mechanism of Schiff bases, (3) Schiff base reactions, (4) metal complexes of Schiff base, (5) classification of Schiff bases, (6) biological activity of Schiff bases, and (7) usage of Schiff bases.

Keywords: Schiff bases, azomethine, biological activity, metal complexes, formation mechanism

1. Introduction

Compounds that are formed as a result of the nucleophilic addition reaction of aldehydes and ketones with primary amines under suitable conditions and which have carbon-nitrogen double bonds (—CH=N—) in their structure are called Schiff bases. Schiff bases, which were first obtained by the German chemist H. Schiff in 1864 [1], were started to be used as ligands by Pfeiffer in the 1930s [2, 3] (**Figure 1**).

Aldehydes react very easily with primary amines to form Schiff bases, but this process is not so easy for ketones. In order to obtain Schiff bases from ketones, it is necessary to pay attention to factors, such as the choice of catalyst, the appropriate pH range, the selection of a solvent that can form an azeotrope mixture with the water to be formed in the reaction, and the appropriate reaction temperature. The carbon-nitrogen double bond in Schiff bases formed as a result of the reaction of primary

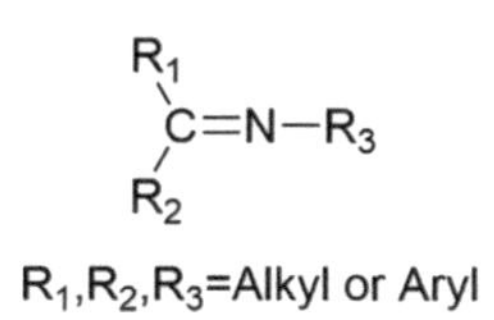

Figure 1.
General representation of Schiff bases.

IntechOpen

amines with aldehydes is called azomethine or aldimine, while the bond formed as a result of reaction with ketone is called imine or ketimine.

Schiff bases are selective toward metal ions and form complexes by transferring electrons from the active ends they contain to the metal. Schiff bases are known as a good nitrogen donor ligand (—CH=N—). During the formation of the coordination compound, one or more electron pairs are donated to the metal ion by these ligands. Schiff bases can form highly stable 4-, 5-, and 6-ring complexes if they donate more than one electron pair. For this, a second functional group with a displaceable hydrogen atom must be found as close as possible to the azomethine group. This group is preferably the hydroxyl group [4].

2. Physical properties of Schiff bases

Schiff bases are usually colored and transparent solids. They are used in the determination of metal amounts and in the identification of carbonyl compounds due to their precise melting points.

The carbon-nitrogen double bond in Schiff bases rotates more easily than the carbon-carbon double bond, which allows stereoisomers to transform into each other. The reason for this: polarization occurs in the azomethine bond due to the fact that nitrogen is more electronegative than carbon (**Figure 2**).

The stereoisomers of Schiff bases cannot be isolated with a few exceptions due to the very small energy difference between them. If only an electronegative group is attached to the nitrogen atom, stereoisomers become isolated, since this group reduces the ease of rotation around the azomethine bond. Since the electronegative group attached to the nitrogen atom in the azomethine group will push the negative charges of the nitrogen atom toward the carbon, this will cause a decrease in polarization and an increase in the character of the covalent double bond.

All compounds containing an azomethine group show basic properties due to the unshared electron pairs on the nitrogen atom and the electron donating feature of the double bond. Schiff bases show weaker basic properties compared to their corresponding amines. The reason for this is that while the nitrogen atom in amines undergoes sp^3 hybridization, this hybridization turns into sp^2 hybridization when the imine structure is formed. Since the s character will increase in hybridization, the basicity will decrease greatly.

The C=N system is a weak chromophore that shows absorption in the ultraviolet field. Conjugation with phenyl groups shifts absorption to the visible region. When there is a deactivating substituent in the aromatic ring, such as a halogen, the wavelength of absorption decreases. Generally, aryl alkyl ketimines are absorbed at values between dialkyl and diaryl ketimines [5]. The IR stretch bands of the C=N system are generally observed at 1610–1635 cm^{-1} and that of $C{=}N^+$ at 1665–1690 cm^{-1} [6].

Figure 2.
Polarization of azomethine bond.

3. Chemical properties of Schiff bases

Schiff bases have many properties that vary according to the substituents attached to the azomethine group. The stability of the azomethine compound increases when there is an electronegative group attached to the nitrogen atom. The best example of this is that oximes carrying hydroxyl groups on the nitrogen atom along with phenylhydrazone and semicarbazones carrying —NH groups are much more stable to hydrolysis than Schiff bases carrying alkyl or aryl substituents on the nitrogen atom. Although Schiff bases are stable against alkalies, they are separated into amine and carbonyl compounds by hydrolysis in acidic environment.

The Schiff base formation reaction is reversible. As a result of the reaction, one mole of water is formed and the water in the environment shifts the direction of the reaction to the left. Therefore, the reaction is usually carried out in solvents where water can be removed from the environment by distillation, forming an azeotrope. If the reaction is carried out using amines containing an electronegative atom with unpaired electrons in the nitrogen atom, the reaction is completed and since hydrolysis will not occur, Schiff bases can be isolated with high efficiency (**Figure 3**).

The structures of Schiff bases are determined by the tautomeric transformations that occur depending on the polarity of the solvent and the hydrogen bonds that occur in the molecule. The preferred conformation in terms of the stability of Schiff bases is the nonplanar structure seen in **Figure 4.** This conformation has also been confirmed by quantum mechanics calculations [7].

In the studies, it has been reported that there are two types of tautomer forms, phenol-imine and ketone-amine, in Schiff bases obtained by using aldehydes containing ortho hydroxy group (**Figure 5**). The presence of these two tautomeric structures was determined by spectroscopic methods such as ^{13}CNMR, ^{1}H-NMR, UV-Vis, and X-ray crystallography [8].

In studies with Schiff bases prepared from 2-hydroxy-1-naphthaldehyde and some aromatic and aliphatic amines (ammonia, methylamine, and phenylamine), it has

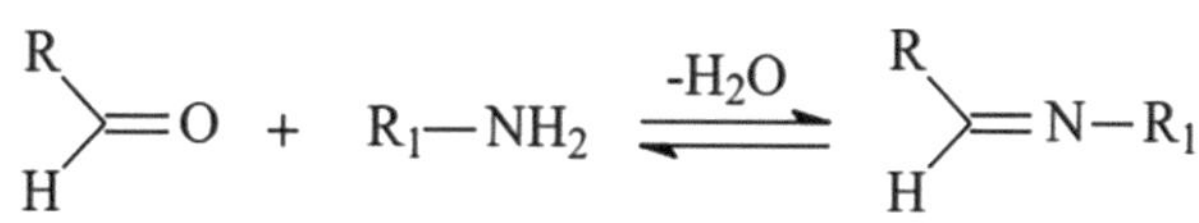

Figure 3.
Schiff base formation reaction.

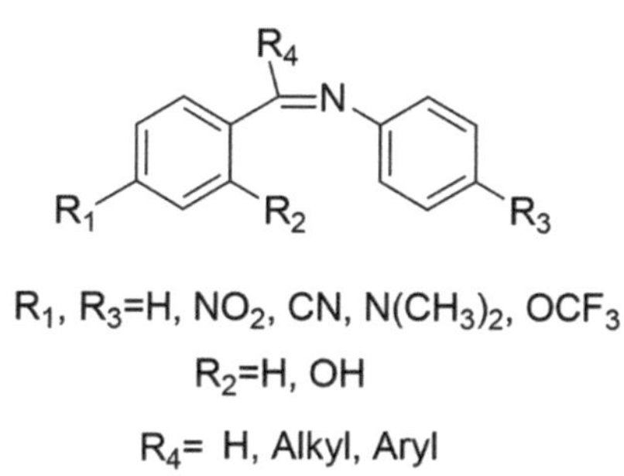

Figure 4.
Preferred conformation of Schiff bases.

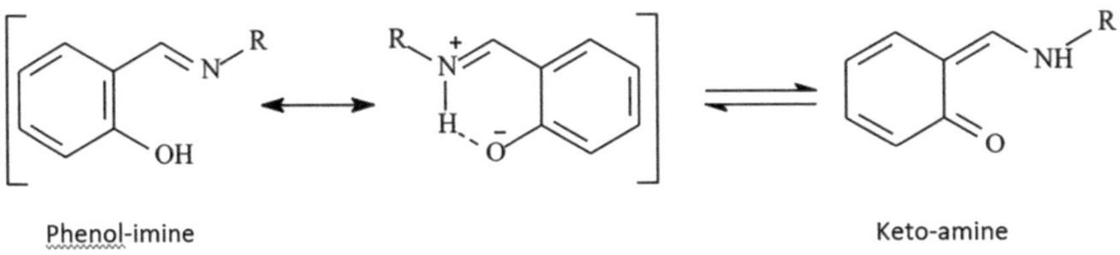

Figure 5.
Tautomeric structure of Schiff bases.

been observed that the keto form is dominant in polar solvents, such as chloroform, and the enol form is dominant in nonpolar solvents [9, 10].

4. Formation mechanism of Schiff bases

The most widely used method discovered by Schiff for the preparation of Schiff bases is the reaction of aliphatic or aromatic aldehydes or ketones with aliphatic or aromatic primary amines. The synthesis of Schiff bases obtained from the reaction of carbonyl compounds with primary amines takes place in two main steps. In the first step, a carbinolamine intermediate is formed from the condensation of the carbonyl group with the primary amine, and in the second step, a Schiff base is formed from the dehydration of the intermediate seen in **Figure 6** [12–14].

The formation of Schiff base is a pH-dependent reaction. Since the amine will form salt at low pH, the free amine concentration decreases and the fast addition step slows down and becomes the step that determines the rate of the reaction mechanism (**Figure 7**). In the case of a decrease in acidity, the addition step is faster and the

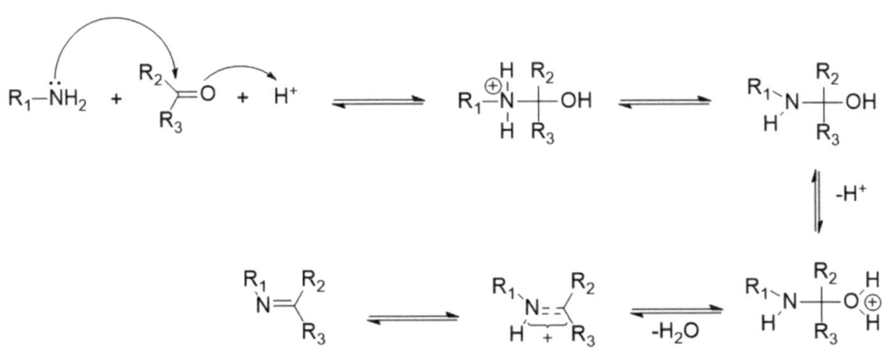

Figure 6.
Mechanism of condensation of carbonyl compounds with amines [11].

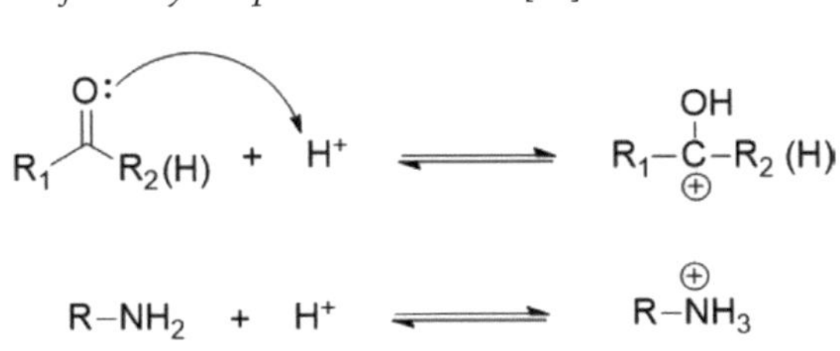

Figure 7.
Increase in electrophilic power and decrease in nucleophilic power in acidic medium.

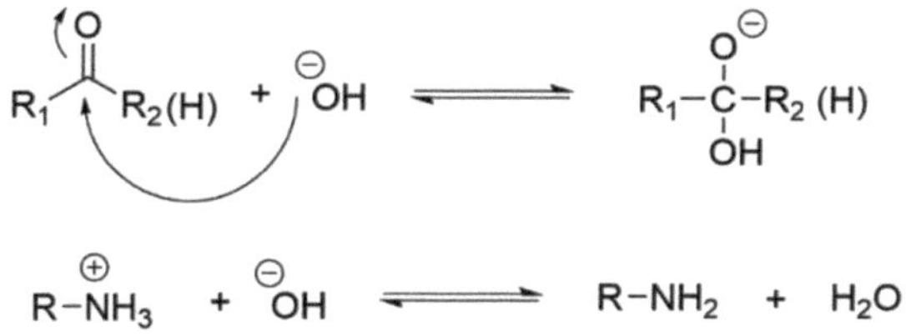

Figure 8.
Decrease in electrophilic power and increase in nucleophilic power in basic medium.

elimination step is slower (**Figure 8**). The optimal pH is the pH between these two extremes (pH 3–4). This pH is suitable for both starting the nucleophilic addition reaction and performing elimination reaction at a sufficient speed [7].

The effect of substitution is great on the stability of Schiff bases. Since small-molecular-weight aliphatic imines without substituents on the nitrogen atom are easily polymerized, detailed information about these imines is not available. Schiff bases containing aryl substituents can be synthesized more stable and easily due to the electron feeding of the imine bond through ring conjugation, while those containing alkyl substituents are relatively unstable, synthesized in a long time, and polymerization is observed.

In the formation of imine; aldehydes are more reactive than ketones because they are less sterically hindered. In addition, in ketones, groups attached to the carbonyl carbon donate electrons, reducing the electrophilic character of the carbonyl carbon, thus reducing the reaction tendency and causing the reaction to take place more slowly. Therefore, although aldehydes and primary amines can easily form Schiff bases, it is quite difficult to obtain Schiff bases from ketones. In order to obtain Schiff base from ketones, many factors, such as choosing a solvent that can form an azeotrope mixture with the water released during the reaction, choosing a catalyst, choosing the appropriate pH range and the appropriate reaction temperature, must be taken into account. Particularly in order to obtain Schiff base from aromatic ketones, high temperature, long reaction time, and catalyst are required [4, 15].

Aromatic aldehydes and ketones can form highly stable Schiff bases. Aromatic aldehydes react with amines at low temperature and in a suitable solvent environment. In the reaction of aromatic aldehydes with aromatic amines, it has been indicated that the reaction rate increases in the presence of an electron-withdrawing substituent in the para position of the aldehyde and decreases in the presence of para position of the amine. While the water formed in the reaction must be removed during the production of Schiff base from aromatic ketones, there is no need to remove water in the synthesis of Schiff base from aldehydes and dialkyl ketones. While the water formed in the reaction must be removed during the generation of Schiff base from aromatic ketones, there is no need to remove water in the synthesis of Schiff base from aldehydes and dialkyl ketones [16].

5. Synthesis methods of Schiff bases

5.1 Reaction of aldehydes and ketones with primary amines

The reaction of primary amines with carbonyl compounds is usually carried out by reflux. Since the reaction is reversible, the water formed in the reaction medium must be removed to prevent hydrolysis. Dean-Stark apparatus is generally used to remove water. In addition, the reaction was carried out successfully by using dehydration agents such as sodium sulfate and molecular sieve [17]. Moreover, methods using solvents, such as tetramethyl orthosilicate or trimethyl orthoformate, which remove water in the reaction medium, have also been reported in the literature [18, 19].

The reaction can be accelerated by acid catalysis. In such cases, mineral acids, such as H_2SO_4 or HCl, organic acids, such as p-toluene sulfonic acid, pyridinium p-toluenesulfonate, acidic resin, montmorillonite, or Lewis acids ($ZnCl_2$, $TiCl_4$, $SnCl_4$, BF_3Et_2O, $Mg(ClO_4)_2$, $MgSO_4$), can be used [15, 20–25].

The reaction of aliphatic ketones with amines to form a Schiff base occurs more slowly than with aldehydes. When the reaction rates of the same primary amine and aldehydes and ketones are compared, it was found that the rate order was; the rate order is aromatic aldehyde>aliphatic aldehyde>aliphatic ketone>aromatic ketone [26]. Recently, new solvent-free techniques have been developed for imine formation, including clay, microwave irradiation, water suspension media, liquid crystal, molecular sieve, and infrared and ultrasonic irradiation [27–34].

5.2 Reaction of organometallic compounds with nitriles

Grignard reagents can react with nitriles to form ketimines. Anhydrous hydrogen chloride or anhydrous ammonia is added to the reaction medium to prevent the hydrolysis of the intermediate products into ketones. With this method, intermediate products can be isolated with an efficiency of 50–90% (**Figure 9**) [26].

5.3 Reaction of phenols and phenol ethers with nitriles

The alkyl or aryl nitriles react with phenol and phenol ethers with high efficiency under acid catalysis to form ketimines [35]. The reaction is carried out by saturating a solution of nitrile and phenol dissolved in ether with HCl gas. $ZnCl_2$ should be used in reactions with lower reactivity phenols (**Figure 10**).

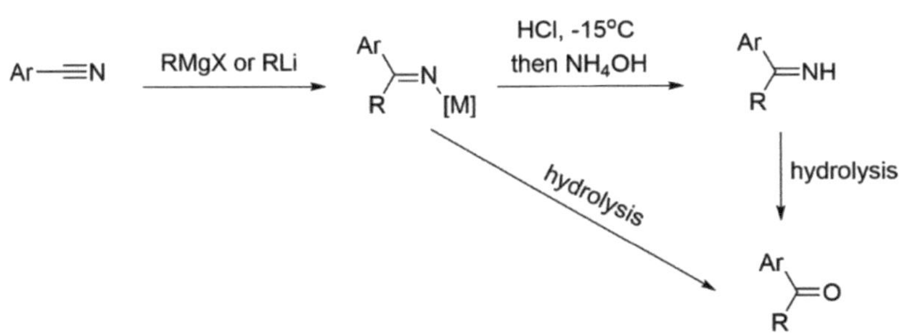

Figure 9.
Addition of organometallic reagents to nitriles.

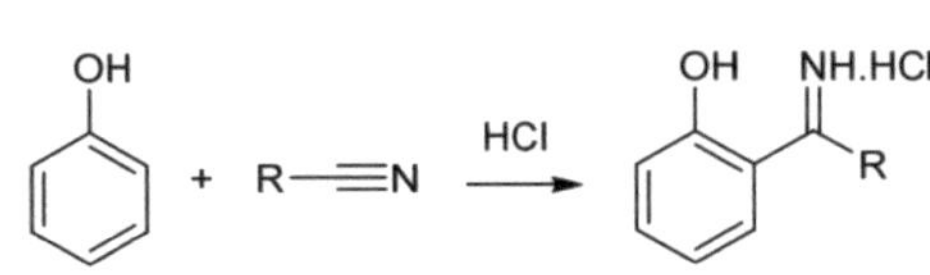

Figure 10.
Reaction of phenols with nitriles.

5.4 Aerobic oxidative synthesis method

Since aldehydes and ketones can be obtained from their corresponding alcohols by oxidative methods, it is also possible to prepare imines from alcohols and amines using oxidative processes (**Figures 11** and **12**) [36–43].

Following this general approach, Huang and Largeron developed new catalytic processes that convert primary and secondary amines to imines by aerobic oxidation under mild conditions [38–44].

5.5 Reaction of metal amides

Calcium or alkali metal salts of primary amines react with aromatic ketones to form Schiff bases [26].

6. Reactions of Schiff bases

6.1 Polymerization reaction

Many studies have been carried out on poly (Schiff bases) over time due to their thermal, conductive [45–47], fiber forming [48], liquid crystal [49, 50], and nonlinear optical properties [51, 52]. One of them is poly (Schiff base) formed by the reaction of diamines and dialdehydes by Catanescu et al. (**Figure 13**) [53].

6.2 Reaction with Zn and haloesters

β-Lactams are formed as a result of the reaction of Schiff bases with Zn and haloesters at room temperature [54].

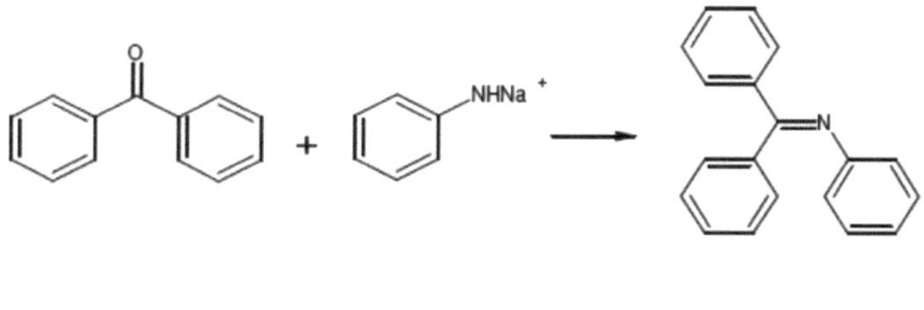

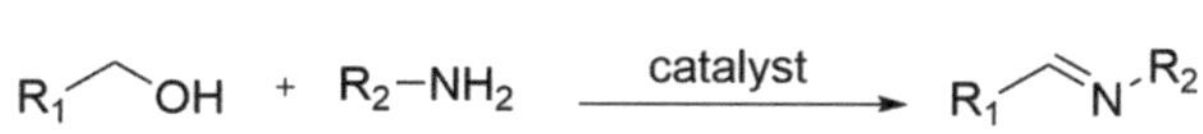

Figure 11.
Oxidative synthesis of imines from alcohols and amines.

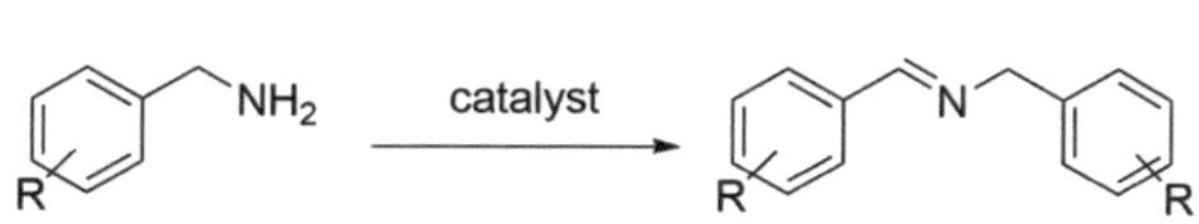

Figure 12.
Oxidative synthesis of imines from amines.

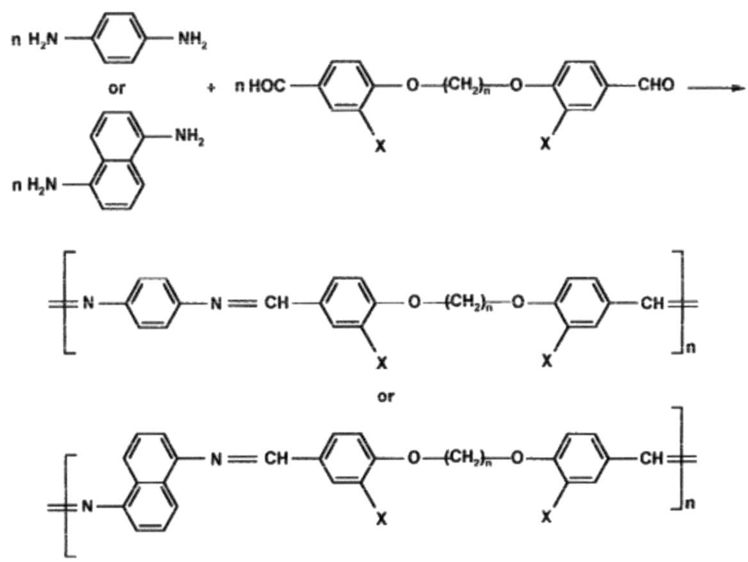

Figure 13.
Polymer synthesis.

6.3 Reaction with HCN

Nitrile derivatives are formed from the reaction of Schiff bases with HCN, and α-amino acids are formed by their hydrolysis [54].

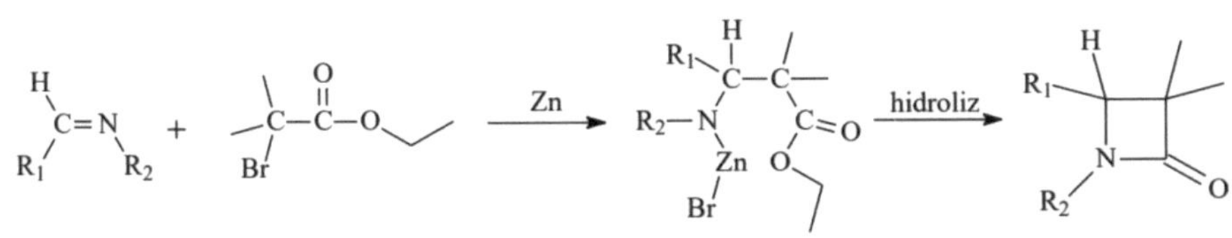

6.4 Reduction reactions

Schiff bases are reduced with $LiAlH_4$, $NaBH_4$, and Na-EtOH reagents to form secondary amines [54].

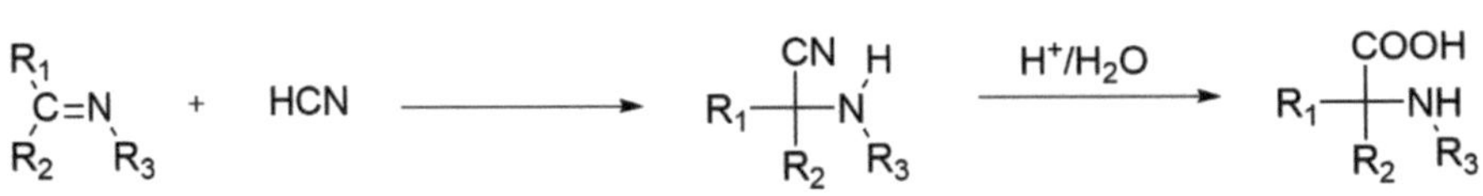

6.5 Hydrolysis

Since the reaction steps of Schiff bases synthesized with carbonyl compounds and amines are reversible, starting materials are obtained by hydrolysis of Schiff bases. In the first step of hydrolysis, the intermediate product, carbinolamine, is formed. In the second step, the carbinolamine is decomposed to form the reaction products aldehyde (or ketone) and amine. Hydrolysis reactions are mostly acid-catalyzed and the rate of the reaction depends on the acidity strength [5].

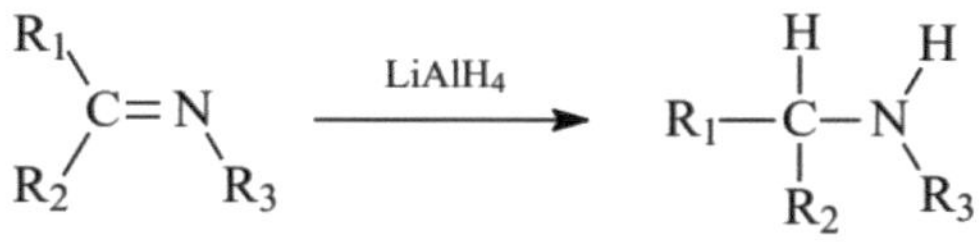

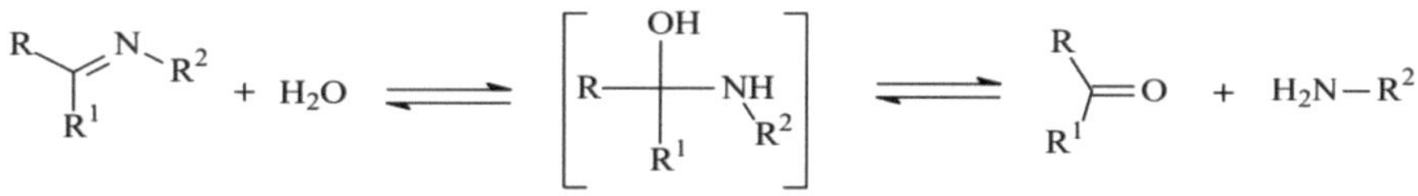

7. Schiff base metal complexes

Schiff bases are widely used as ligands in coordination chemistry. Ligands are called Lewis bases because they donate electron pairs to the central atom. Since the nitrogen atom in the imine bond contains unpaired electrons, Schiff bases are electron donors, basic in character, and can form complexes with almost all transition metals. This atom, also known as azomethine nitrogen, is the primary bonding point for the Schiff base. In addition, the azomethine system, where the nitrogen atom is bonded with a double bond, can be a coordination site for d-metal ions suitable for back bonding, by means of its π-orbitals. Thus, the azomethine group with the nitrogen atom has both σ-donor and π-acceptor functions. This provides high stability for metal complexes formed by Schiff bases. Schiff bases can form stable compounds with metal ions if they have a structure that can form a quintet or hexavalent chelate ring [55].

Apart from the fact that the azomethine group is basic, there must be a functional group in the molecule close to the azomethine bond, from which the hydrogen atom can be easily removed, in order to form stable complexes as a ligand. Since Schiff base ligands, which have groups such as OH and SH, in the ortho position adjacent to the azomethine bond, form a six-membered ring with the metal, stable complexes are formed [56].

The properties of complex compounds vary depending on the ligand and metal ion used. The size, charge, and ionization potential of the metal ion used in the formation of the complex affect the stability of the complex. Since the substituents can change the basicity of the Schiff base imine nitrogen, the ligand property also changes depending on the substituents. Therefore, the stability of the metal complexes of Schiff bases is more or less affected by the substituents in their structure.

Three methods are generally used to synthesize metal complexes of Schiff bases. These methods are direct interaction of metal salt with Schiff base [54]; condensation of aldehyde, amine, and metal salt with the effect of template [57]; and condensation of aldehyde complexes with amines [58].

The first studies on Schiff base metal complexes were made by spectrophotometric techniques [59]. Later, potentiometric studies were started by Leussing et al. When these studies were examined, it was seen that Schiff bases formed complexes with metal ions in 1:1 and 1:2 ratios [60].

7.1 Classification of Schiff base complexes

Classification of Schiff bases metal complexes is done by considering the donor atoms of the compound. Depending on the type and number of donor atoms they contain, some of the metal complexes encountered are those with N—O, O—N—O, O—N—S, N—N—O, O—N—N—O, and N—N—N—N donor atom systems.

7.1.1 N—O Type Schiff base complexes

N—O type Schiff base formed by salicylaldehyde and p-N, N′-dimethylaniline is bidentate and forms a 1:1 complex with Ag^+ ion (**Figure 14**) [61].

7.1.2 O—N—O type Schiff base complexes

$[ML]X_2$ type colored complexes synthesized with (E)–N′-((7-hydroxy-4-methyl-2-oxo-2H-chromen-8-yl)methylene)benzohydrazide ligand are examples of O—N—O type Schiff base complexes (**Figure 15**). This ligand forms a complex with metal ions by reacting at a ratio of 2:1. M = Mn(II), Co(II), Ni(II), Cu(II), Sr(II) Cd(II), X = Cl^- [62].

7.1.3 O—N—S type Schiff base complexes

Schiff base, which consists of 2-hydroxy-1-naphthaldehyde and 2-aminoetantiol, can be given as an example of such complexes. These complexes are tridentate and they have dibasic properties. The structure of the UO_2^{6+}(VI) complex of this Schiff base is shown in **Figure 16** [63].

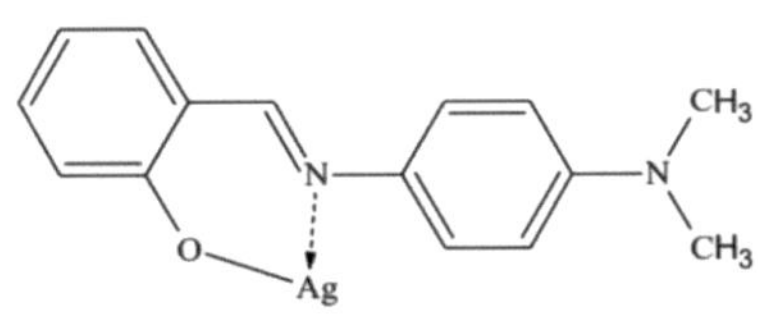

Figure 14.
N—O type Schiff base metal complex.

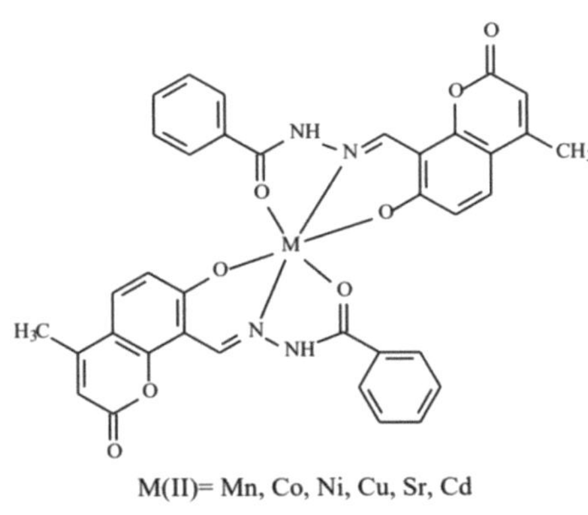

Figure 15.
O—N—O type Schiff base metal complex.

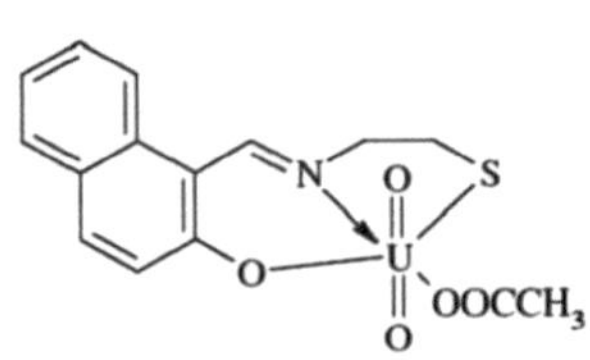

Figure 16.
O—N—S type Schiff base metal complex.

7.1.4 N—N—O type Schiff base complexes

Co(II), Ni(II), and Cu(II) complexes of 4-Chloromethyl-2-(2-hydroxynaphthylidenehydrazine) thiazole ligand can be given as examples of such complexes [64] (**Figure 17**).

7.1.5 O—N—N-O type Schiff base complexes

An example of this group is the oxovanadium(IV) complex synthesized with the N,N′-bis(salicylidine)ethylenediamine ligand [65]. N,N′-bis(salicylidine) ethylenediamine Schiff base, which is obtained with the condensation reaction of salicylaldehyde and ethylenediamine, is commonly known as salen and is an ONNO type tetradentate ligand (**Figure 18**) [66].

7.1.6 N—N—N—N type Schiff base complexes

An example of such complexes is the Fe^{+2} complex of N,N′-bis(2-salicylideniminobenzoyl) ethylenediamine obtained from N,N′-bis(2-aminobenzoyl) ethylenediamine and salicylaldehyde (**Figure 19**) [61].

8. Biological activity of Schiff bases

Schiff bases and metal complexes are vital compounds for biological systems because they have anticancer [67, 68], antibacterial [69–74], antifungal [75, 76], and antiviral [77] properties.

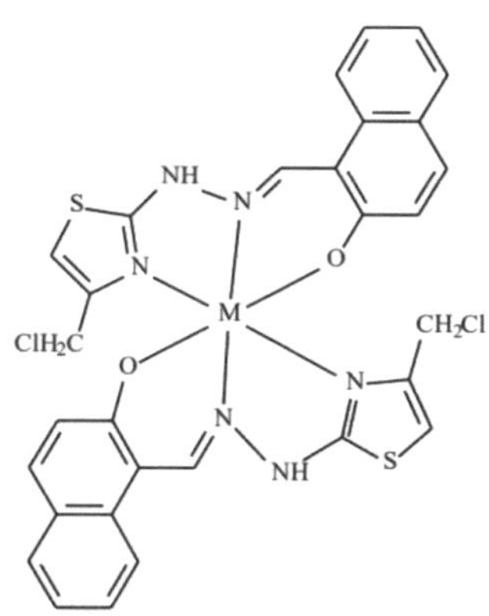

Figure 17.
N—N—O type Schiff base metal complex.

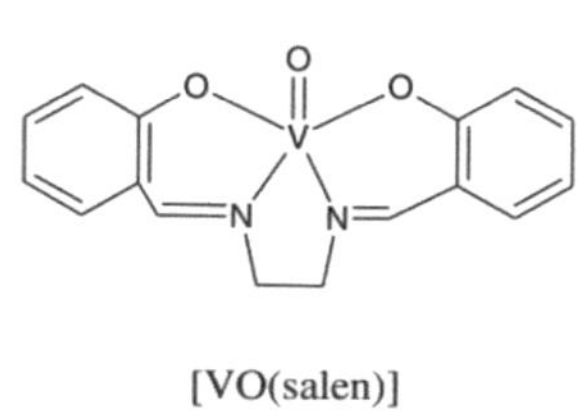

Figure 18.
O—N—N—O type Schiff base metal complex.

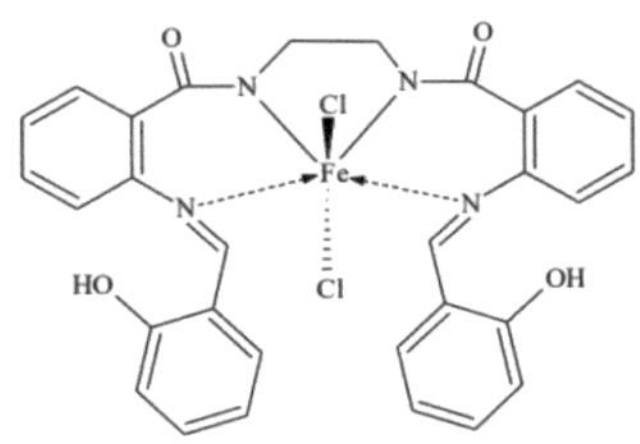

Figure 19.
N—N—N—N type Schiff base metal complex.

The most important biological activity of Schiff bases is the role they play in amino acid biosynthesis, one of the basic processes of life. Schiff bases are important intermediates formed in the biosynthesis of α-amino acids ($RCHNH_2COOH$) used in the synthesis of proteins in organisms. If there is not enough essential amino acid in the food, the organism converts an excess amino acid to the amino acid it needs by transamination reaction in some cases. In this process, the amino group of the excess amino acid is transferred to the keto acid *via* a series of Schiff bases (**Figure 20**) [78].

In addition, Schiff bases have been proven to be present in a variety of natural, semisynthetic, and synthetic compounds and are essential for their biological activity (**Figure 21**) [26].

9. The importance and uses of Schiff bases

Today, with the development of spectral methods, the mechanisms of biochemical reactions can be explained. It has been determined that some reactions in living organisms proceed on Schiff bases, Schiff bases eliminate the toxic effects of aldehyde and amine components and bind free metal ions. The importance of Schiff bases increases day by day as they can be used in biological systems [79], chemical catalysis [80], medicine and pharmacy [81], chemical analysis, and new technologies [82, 83]

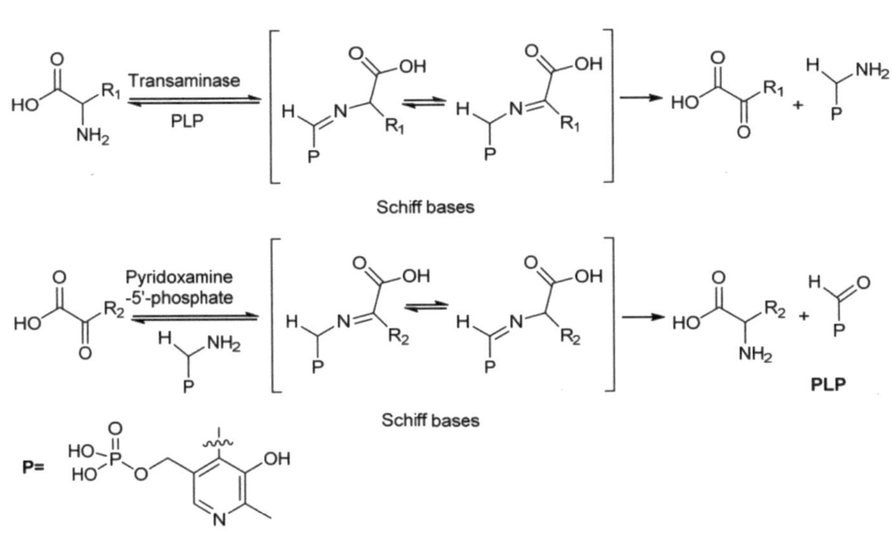

Figure 20.
Transamination reaction through Schiff bases from amino acid to keto acid and vice versa [26].

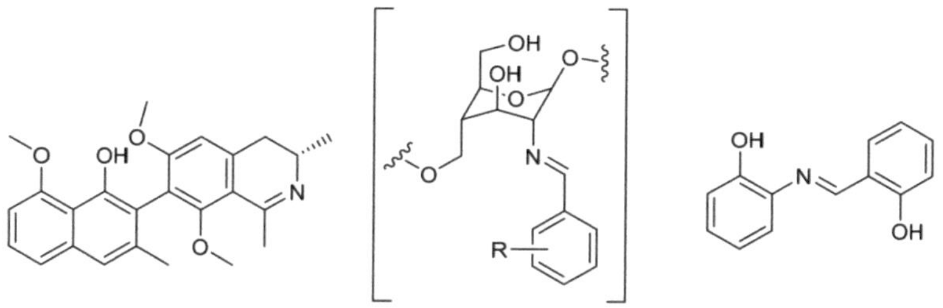

Ancistrocladidine
(Antimalarial activity)

Chitosan-derived Schiff base
R = H or OH
(Antifungal activity)

***N*-(Salicylidene)-2-hydroxyaniline (71)**
(Antibacterial activity)

Figure 21.
Examples of biologically active Schiff bases.

due to their properties. If we briefly summarize the reasons why Schiff bases have such a wide working area [84]:

- Alternatively, Schiff bases can be synthesized by the template effect. This procedure directly gives the designed complexes. Also, these complexes can give transmetalation reactions when reacted with a different metal salt. This method ensures that complexes that cannot be obtained by different methods are obtained with sufficient purity and high yield.

- Since Schiff bases generally contain additional donor groups, such as N, O, S, and P, they can form stable complexes with almost all metals. In addition to these features, they play important roles in biological systems.

- Schiff bases are easily synthesized using carbonyl compounds and primary amines and can be functionalized in a wide variety of ways using appropriate groups.

- When treated with suitable reducing agents, they can form polyamine derivatives that are more flexible and less sensitive to hydrolysis. These reduced compounds contain NH groups that can be further functionalized by suitable synthetic procedures.

- By bonding to a suitable support, such as silica, they result in modified catalysts and different designed surfaces.

- By binding groups, such as crown ether, macrocyclic thioether, and polyaza derivatives, special ligands can be formed to form selective systems with the capacity to bind different metal ions.

- Due to the interaction of metal chelates with the DNA helix, they can be used in the design of new models in diagnosis and therapy.

In addition, due to their photochromism feature, they can be used in different fields, such as controlling and measuring radiation intensity, image systems, and optical computers [85]. Since the metal complexes of these compounds are colored substances, they are used as pigment dyestuffs in the dye industry, especially in textile dyeing [86, 87]. Moreover, they are widely used in the perfume and pharmaceutical industries. These compounds also have properties, such as synthetic oxygen carrier and intermediate in enzymatic reactions. They are also used as spectrophotometric reagents in analytical chemistry, as they react selectively and specifically to some metal ions. They can be used in aircraft construction, television and computer screens, and digital clock displays by taking advantage of the liquid crystal feature that occurs in some metal complexes [88].

10. Application of Schiff bases in organic synthesis

We can classify the reactions in which Schiff bases are used as precursors in four groups as seen in **Figure 22** [26].

I. Addition of organometallic reagent or hydride to C=N bond to form asymmetric C-C bond [89, 90].

II. Hetero Diels-Alder reaction to produce six-membered nitrogen-containing heterocyclic compounds [91–93].

III. Use of chiral salen metal complexes in the asymmetric synthesis [94–96].

IV. Staudinger reaction with ketene to provide biologically important β-lactams [97–99].

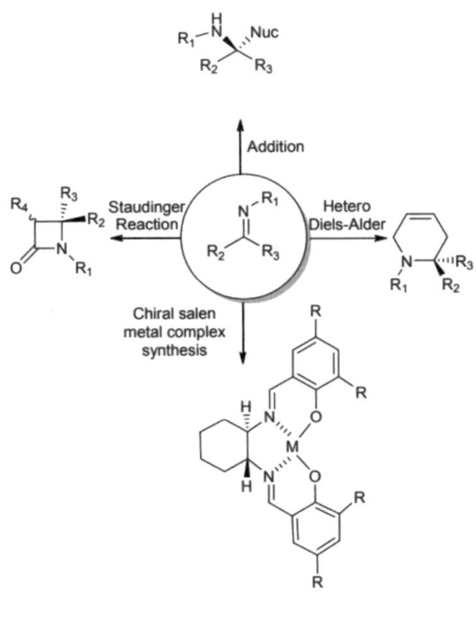

Figure 22.
Applications of Schiff bases as a starting material in organic synthesis.

Author details

Nuriye Tuna Subasi
Department of Food Engineering, Ahi Evran University, Kırşehir, Turkey

*Address all correspondence to: tunasubasi@gmail.com

References

[1] Schiff H. Untersuchungen Uber salicin derivate. European Journal of Organic Chemistry. 1869;**150**:193-200

[2] Pfeiffer P, Breith E, Lubbe E, Tsumaki T. Thcylische Orthokondesierte Nebenvalenzringe. Annalen Der Chemie. 1932;**492**:81-127

[3] Pfeiffer P, Breith E, Lubbe E, Tsumaki T. Thcylische Orthokondezierte Nebenvalenzringe. Annalen Der Chemie. 1933;**503**:84-127

[4] Patai S. The Chemistry of the Carbon-Nitrogen Double Bond. Norwich: Wiley; 1970

[5] Layter RW. The chemistry of imines. Chemical Reviews. 1963;**63**:489-510

[6] Smith PAS. The Chemistry of Open-Chain Organic Nitrogen Compounds. New York: Benjamin; 1965

[7] Ekmekcioglu P. Synthesis and Characterization of New Six Dentate Schiff Bases with Cu(II), Ni(II) and Co(II) Complexes. Trabzon: Karadeniz Technical University; 2015

[8] Metzler DE, Christen P. Transaminases. New York: John Wiley; 1985

[9] Dudek GO, Dudek EP. Spestrocopic Studies of Keto-Enol Equilibria. VII. Nitrogen-15 Substitued Schiff Bases. Journal of the American Chemical Society. 1964;**86**(209):4283-4287

[10] Dudek GO, Dudek EP. Spectroscopic Studies of Keto-Enol Equilibria. IX. N15-Substituted Anilides. Journal of the American Chemical Society. 1966; **88**(11):2407-2412

[11] Borisova NE, Reshetova MD, Ustynyuk YA. Metal-free methods in the synthesis of macrocyclic Schiff bases. Chemical Reviews. 2007;**107**:46-79

[12] Schiff H. Eine neue Reihe organischer Diamine (in German). Justus Liebigs Annalen der Chemie. 1866;**140**:92-137

[13] Robertson GM. Imines and their N-substituted derivatives: NH, NR and N-haloimines. In: Katritzky AR, Meth-Cohn O, Rees CW, editors. Comprehensive Organic Functional Group Transformations. 1st ed. Amsterdam, The Netherlands: Elsevier; 1995. pp. 403-423

[14] Dobbs AP, Rossiter S. Imines and their N-substituted derivatives: NH, NR, and N-Haloimines. In: Alan RK, Taylor RJK, editors. Comprehensive Organic Functional Group Transformations II. Oxford, UK: Elsevier; 2005. pp. 419-450

[15] Billman JH, Tai KM. Reduction of Schiff Bases. II. Benzhydrylamines structurally related compounds. The Journal of Organic Chemistry. 1958; **23**(4):535-539

[16] Pratt EF, Kamet MJ. Reaction rates by distillation. IX. The condensation of anilines with benzaldehydes. The Journal of Organic Chemistry. 1961;**26**(10): 4029-4031

[17] Westheimer FH, Taguchi K. Catalysis by molecular sieves in the preparation of ketimines and enamines. The Journal of Organic Chemistry. 1971; **36**:1570-1572

[18] Love BE, Ren J. Synthesis of sterically hindered imines. The Journal of Organic Chemistry. 1993;**58**:5556-5557

[19] Look GC, Murphy MM, Campbell DA, Gallop MA.

Trimethylorthoformate: A mild and effective dehydrating reagent for solution and solid phase imine formation. Tetrahedron Letters. 1995;**36**: 2937-2940

[20] White WA, Weingarten H. A versatile new enamine synthesis. The Journal of Organic Chemistry. 1967;**32**: 213-214

[21] Liu G, Cogan DA, Owens TD, Tang TP, Ellman JA. Synthesis of enantiomerically Pure N-tert-butanesulfinyl imines (tert-butanesulfinimines) by the direct condensation of tert-butanesulfinamide with aldehydes and ketones. The Journal of Organic Chemistry. 1999;**64**: 1278-1284

[22] Chakraborti AK, Bhagat S, Rudrawar S. Magnesium perchlorate as an efficient catalyst for the synthesis of imines and phenylhydrazones. Tetrahedron Letters. 2004;**45**:7641-7644

[23] Panneerselvam P, Nair RR, Vijayalakshmi G, Subramanian EH, Sridhar SK. Synthesis of Schiff bases of 4-(4aminophenyl)-morpholine as potential antimicrobial agents. European Journal of Medicinal Chemistry. 2005; **40**:225-229

[24] Dalpozzo R, de Nino A, Nardi M, Russo B, Procopio A. Erbium(III) triflate: A valuable catalyst for the synthesis of aldimines, ketimines and enaminones. Synthesis. 2006;7:1127-1132

[25] Naeimi H, Salimi F, Rabiei K. Mild and convenient one pot synthesis of Schiff bases in the presence of P_2O_5/Al_2O_3 as new catalyst under solvent-free conditions. Journal of Molecular Catalysis A: Chemical. 2006;**260**:100-104

[26] Qin W, Long S, Panunzio M, Biondi S. Review Schiff Bases: A short survey on an evergreen chemistry tool. Molecules. 2013;**18**:12264-12289

[27] Varma RS, Dahiya R, Kumar S. Clay catalyzed synthesis of imines and enamines under solvent-free conditions using microwave irradiation. Tetrahedron Letters. 1997;**38**:2039-2042

[28] Schmeyers J, Toda F, Boy J, Kaupp G. Quantitative solid–solid synthesis of azomethines. Journal of Chemical Society. 1998;**4**:989-994

[29] Vass A, Duda SJ, Varma RS. Solvent-free synthesis of Nsulfonylimines using microwave irradiation. Tetrahedron Letters. 1999;**40**:4951-4954

[30] Tanaka K, Shiraishi R. Clean and efficient condensation reactions of aldehydes and amines in a water suspension medium. Green Chemistry. 2000;**2**:272-273

[31] Andrade CKZ, Takada SCS, Alves LM, Rodrigues JP, Suarez PAZ, Brandão RF, et al. Molecular sieves in ionic liquids as an efficient and recyclable medium for the synthesis of imines. Synlett. 2004;**12**:2135-2138

[32] Vázquez MÁ, Landa M, Reyes L, Miranda R, Tamariz J, Delgado F. Infrared irradiation: Effective promoter in the formation of N-benzylideneanilines in the absence of solvent. Synthetic Communications. 2004;**34**:2705-2718

[33] Gopalakrishnan M, Sureshkumar P, Kanagarajan V, Thanusu J. New environmentally-friendly solvent-free synthesis of imines using calcium oxide under microwave irradiation. Research on Chemical Intermediates. 2007;**33**: 541-548

[34] Guzen KP, Guarezemini AS, Órfão ATG, Cella R, Pereira CMP,

Stefani HA. Eco-friendly synthesis of imines by ultrasound irradiation. Tetrahedron Letters. 2007;**48**:1845-1848

[35] Houben J, Fischer W. Formation of aromatic nitriles by basic hydrolysis of trichloromethyl aryl ketimines. Acidic hydrolysis yields ketones. Journal für Praktische Chemie. 1929;**123**:262-313

[36] Shiraishi Y, Ikeda M, Tsukamoto D, Tanaka S, Hirai T. One-pot synthesis of imines from alcohols and amines with TiO_2 loading Pt nanoparticles under UV irradiation. Chemical Communications. 2011;**47**:4811-4813

[37] Jiang L, Jin L, Tian H, Yuan X, Yu X, Xu Q. Direct and mild palladium-catalyzed aerobic oxidative synthesis of imines from alcohols and amines under ambient conditions. Chemical Communications. 2011;**47**:10833-10835

[38] Huang B, Tian H, Lin S, Xie M, Yu X, Xu Q. Cu(I)/TEMPO-catalyzed aerobic oxidative synthesis of imines directly from primary and secondary amines under ambient and neat conditions. Tetrahedron Letters. 2013;**54**:2861-2864

[39] Soule JF, Miyamura H, Kobayashi S. Selective imine formation from alcohols and amines catalyzed by polymer incarcerated gold/palladium alloy nanoparticles with molecular oxygen as an oxidant. Chemical Communications. 2013;**49**:355-357

[40] Gnanaprakasam B, Zhang J, Milstein D. Direct synthesis of imines from alcohols and amines with liberation of H_2. Angewandte Chemie, International Edition. 2010;**49**: 1468-1471

[41] Yuan H, Yoo WJ, Miyamura H, Kobayashi S. Discovery of a metalloenzyme-like cooperative catalytic system of metal nanoclusters and catechol derivatives for the aerobic oxidation of amines. Journal of the American Chemical Society. 2012;**134**: 13970-13973

[42] Largeron M, Fleury MB. Bioinspired oxidation catalysts. Science. 2013;**339**: 43-44

[43] Lan YS, Liao BS, Liu YH, Peng SM, Liu ST. Preparation of imines by oxidative coupling of benzyl alcohols with amines catalysed by dicopper complexes. European Journal of Organic Chemistry. 2013;**23**:5160-5164

[44] Largeron M. Protocols for the catalytic oxidation of primary amines to imines. European Journal of Organic Chemistry. 2013;**24**:5225-5235

[45] Yang CJ, Jenekhe SA. Conjugated aromatic polyimines. 2. Synthesis, structure, and properties of new aromatic polyazomethines. Macromolecules. 1995;**28**(4):1180-1196

[46] Morgan PW, Kwolek SL, Pletcher TC. Aromatic azomethine polymers and fibers. Macromolecules. 1987;**20**(4):729-739

[47] Destri S, Porzio W, Dubitsky Y. Design preparation and properties of regularly alternating conjugated low molecular weight copolymers based on thienylene-azomethine moietics. Synthetic Metals. 1995;**75**(1):25-36

[48] Yang HH. Aromatic High-strength Fibers. New York: Wiley-Interscience; 1989

[49] Cerrada P, Oriol L, Piñol M, Serrano JL, Alonso PJ, Puértolas JA, et al. Influence of hydroxy functionalization and metal cross-linking on fiber properties of liquid-crystalline polyazomethines. Macromolecules. 1999; **32**:3565-3573

[50] Ogiri S, Ikeda M, Kanazawa A, Shiono T, Ikeda T. Polymerization of liquid-crystalline monomers having a Schiff-base structure. Polymer. 1999;**40**: 2145-2150

[51] Thomas O, Inganas O, Anderson MR. Synthesis and properties of a soluble conjugated poly(azomethine) with high molecular weight. Macromolecules. 1998;**31**(8):2676-2678

[52] Weaver MS, Bradley DDC. Organic electroluminescence devices fabricated with chemical vapour deposited polyazomethine films. Synthetic Metals. 1996;**83**:61-66

[53] Catanescu O, Grigoras M, Colotin G, Dobreanu A, Hurduc N, Simionescu CI. Synthesis and Characterization of Some Alifatik-aromatic Poly(Schiff base)s. European Polymer Journal. 2001;**37**: 2213-2216

[54] March J. Advanced Organic Chemistry. NY, USA: Wiley; 1972

[55] Öztürk NS. Synthesis Schiff Bases Derived from Some Anilines and Different Pyridine Aldehydes and Preparing Their Some Transition Metal Complexes. İstanbul: İstanbul University; 1998

[56] Ancin N, İde S, Özta SG, Tüzün M. Spectral and structural studies of Ni(II) and Zn(II) Complexes of N-Trans-Cinnamylidene-2-Mercaptoaniline. Journal of Molecular Structure. 2002; **606**:45-50

[57] Fernández-G JM, del Rio-Portillo F, Quiroz-García B, Toscano RA, Salcedo R. The structures of some orto-hydroxy Schiff Base ligands. Journal of Molecular Structure. 2001;**561**:197-207

[58] Krygowski TM, Woźniak K, Anulewicz R, Pawlak D, Kolodziejski W, Grech E, et al. Through-resonance ionic hydrogen bonding in 5-nitro-nsalicylideneethylamine. The Journal of Physical Chemistry. A. 1997;**101**: 9399-9404

[59] Metzler CM, Cahill A, Metzler DE. Equilibriums and absorption spectra of Schiff bases. Journal of the American Chemical Society. 1980;**102**(19): 6075-6082

[60] Helmut S. Metal Ions in Biological Systems. New York and Basel: CRC Press; 1979. p. 304

[61] Sahal H. Synthesis and Characterization of N-O Type Schiff Base Ligands and Investigation of Their Transition Metal Complexes. Elazığ: Fırat University; 2011

[62] Naik KHK, Selvaraj S, Naik N. Metal complexes of ONO donor Schiff base ligand as a new class of bioactive compounds; Synthesis, characterization and biological evolution. Spectrochimica Acta Part A. 2014;**131**:599-605

[63] Syamal A, Singhal OP. Synthesis and characterization of new dioxouranium (VI) complexes with tridentate sulfur donor ligands. Journal of Inorganic and Nuclear Chemistry. 1981;**43**(11):2821-2825

[64] Saydam S, Yılmaz E. Synthesis, characterization, and thermal behavior of 4- Chloromethyl-2-(2-Hydroxynaphthylidenehydrazino) Thiazole and its Complexes with Cr(III), Co(II), Ni(II), and Cu(II). Synthesis and Reactivity in Inorganic, Metal-Organic, and Nano-Metal Chemistry. 2005; **35**(10):767-771

[65] Bialek M, Leksza A, Piechota A, Kurzak K, Koprek K. Oxovanadium(IV) complexes with [ONNO]-chelating ligands as catalysts for ethylene homo-

and copolymerization. Journal of Polymer Research. 2014;**21**(4):389

[66] Baleizao C, Garcia H. Chiral Salen complexes: An overview to recoverable and reusable homogeneous and heterogeneous catalysts. Chemical Reviews. 2006;**106**:3987-4043

[67] Desai SB, Desai PB, Desai KR. Synthesis of some Schiff bases, thiazolidinones and azetidinones derived from 2,6-diaminobenzo1,2-d: 4,5-d' bisthiazole and their anticancer activities. Heterocyclic Communications. 2001;7:83-90

[68] Przybylski P, Huczynski A, Pyta K, Brzezinski B, Bartl F. Biological properties of Schiff bases and azo derivatives of phenols. Current Organic Chemistry. 2009;**13**:124-148

[69] Abdel Aziz AA, Salem ANM, Sayed MA, Aboaly MM. Synthesis, structural characterization, thermal studies, catalytic efficiency and antimicrobial activity of some M(II) complexes with ONO tridentate Schiff base N-salicylidene-o-aminophenol (saphH_2). Journal of Molecular Structure. 2012;**1010**:130-138

[70] Sinha D, Tiwari AK, Singh S, Shukla G, Mishra P, Chandra H, et al. Synthesis, characterization and biological activity of Schiff base analogues of indole-3-carboxaldehyde. European Journal of Medicinal Chemistry. 2008;**43**:160-165

[71] Ronad PM, Noolvi MN, Sapkal S, Dharbhamulla S, Maddi VS. Synthesis and antimicrobial activity of 7-(2-substituted phenylthiazolidinyl)-benzopyran-2-one derivatives. European Journal of Medicinal Chemistry. 2010;**45**:85-89

[72] Amin R, Krammer B, Abdel-Kader N, Verwanger T, El-Ansary A. Antibacterial effect of some benzopyrone derivatives. European Journal of Medicinal Chemistry. 2010; **45**:372-378

[73] Karthikeyan MS, Prasad DJ, Poojary B, Bhat KS, Holla BS, Kumari NS. Synthesis and biological activity of Schiff and Mannich bases bearing 2,4-dichloro-5-fluorophenyl moiety. Bioorganic & Medicinal Chemistry. 2006;**14**:7482-7489

[74] Saravanan G, Pannerselvam P, Prakash CR. Synthesis and anti-microbial screening of novel Schiff bases of 3-amino-2-methyl quinazolin 4-(3H)-one. Journal of Advanced Pharmaceutical Technology & Research. 2010;**1**:320-325

[75] De Souza AO, Galetti FCS, Silva CL, Bicalho B, Parma MM, Fonseca SF, et al. Antimycobacterial and cytotoxicity activity of synthetic and natural compounds. Quimica Nova. 2007;**30**: 1563-1566

[76] Rathelot P, Vanelle P, Gasquet M, Delmas F, Crozet MP, Timon-David P, et al. Synthesis of novel functionalized 5-nitroisoquinolines and evaluation of in vitro antimalarial activity. European Journal of Medicinal Chemistry. 1995;**30**: 503-508

[77] Jarrahpour A, Khalili D, de Clercq E, Salmi C, Brunel JM. Synthesis, antibacterial, antifungal and antiviral activity evaluation of some new bis-Schiff bases of isatin and their derivatives. Molecules. 2007;**12**: 1720-1730

[78] Snell EE, Jenkins WT. The mechanism of the transamination reaction. Journal of Cellular and Comparative Physiology. 1959;**54**: 161-177

[79] Sharaby CM, Amine MF, Hamed AA. Synthesis, structure characterization and biological activity of selected metal complexes of sulfonamide Schiff base as a primary ligand and some mixed ligand complexes with glycine as a secondary ligand. Journal of Molecular Structure. 2017; **1134**:208-216

[80] Redshaw C. Use of metal catalysts bearing Schiff Base macrocycles for the ring opening polymerization (ROP) of cyclic esters. Catalysts. 2017;7(5):165-176

[81] Roberts DW, Schultz TW, Api AM. Skin sensitization QMM for HRIPT NOEL data: Aldehyde Schiff-Base domain. Chemical Research in Toxicology. 2017;**30**(6):1309

[82] DiRisio RJ, Armstrong JE, Frank MA, Lake WR, McNamara WR. Cobalt Schiff-base complexes for electrocatalytic hydrogen generation. Dalton Transactions. 2017;**46**: 10418-10425

[83] Upadhyay KK, Kumar A, Upadhyay S, Mishra PC. Synthesis, characterization, structural optimization using density functional theory and superoxide ion scavenging activity of some Schiff bases. Journal of Molecular Structure. 2008;**873**:5-16

[84] Vigato PA, Tamburini S. The challenge of cyclic and acyclic Schiff bases and related derivatives. Coordination Chemistry Reviews. 2004; **248**(17-20):1717-2128

[85] Yeap GY, Ha ST, Ishizawa N, Suda K, Boey PL, Mahmood WAK. Synthesis, crystal structure and spectroscopic study of para substituted 2-hydroxy-3- methoxybenzalideneanilines. Journal of Molecular Structure. 2003;**658**(1-2):87-99

[86] Lasri J, Elsherbiny AS, Eltayeb NE, Haukka M, El-Hefnawy ME. Synthesis and characterization of ferrocene-based Schiff base and ferrocenecarboxaldehyde oxime and their adsorptive removal of methyl blue from aqueous solution. The Journal of Organic Chemistry. 2018;**866**:21-26

[87] Zeishen W, Zigi G, Zhenhuan Y. Synthesis, characterization and anticanser activity of L – alanin Schiff base complexes of cooper(II), zinc(II), an cobalt(II). Synthesis and Reactivity in Inorganic and Metal-Organic Chemistry. 1990;**20**(3):335-344

[88] Zhang J, Xu L, Wong WY. Energy materials based on metal Schiff base complexes. Coordination Chemistry Reviews. 2018;**355**:180-198

[89] Kobayashi S, Mori Y, Fossey JS, Salter MM. Catalytic enantioselective formation of C—C bonds by addition to imines and hydrazones: A ten-year update. Chemical Reviews. 2011;**111**: 2626-2704

[90] Guizzetti S, Benaglia M. Trichlorosilane-mediated stereoselective reduction of C=N bonds. European Journal of Organic Chemistry. 2010;**29**: 5529-5541

[91] Masson G, Lalli C, Benohoud M, Dagousset G. Catalytic enantioselective [4 + 2] cycloaddition: A strategy to access aza-hexacycles. Chemical Society Reviews. 2013;**42**:902-923

[92] Smith CD, Gavrilyuk JI, Lough AJ, Batey RA. Lewis acid catalyzed three-component Hetero-Diels-Alder (Povarov) reaction of N-arylimines with strained norbornene-derived dienophiles. The Journal of Organic Chemistry. 2010;**75**:702-715

[93] Jørgensen KA. Catalytic asymmetric Hetero-Diels-Alder reactions of carbonyl

compounds and imines. Angewate Chemie International Edition. 2000;**39**: 3558-3588

[94] Cozzi PG. Metal-Salen Schiff base complexes in catalysis: Practical aspects. Chemical Society Reviews. 2004;**33**: 410-421

[95] Katsuki T. Unique asymmetric catalysis of cis-beta metal complexes of salen and its related Schiff-base ligands. Chemical Society Reviews. 2004;**33**: 437-444

[96] Matsunaga S, Shibasaki M. Multimetallic schiff base complexes as cooperative asymmetric catalysts. Synthesis. 2013;**45**:421-437

[97] D'hooghe M, van Brabandt W, Dekeukeleire S, Dejaegher Y, de Kimpe N. Highly stereoselective synthesis of beta-lactams utilizing alpha-chloroimines as new and powerful chiral inductors. Chemistry - A European Journal. 2008;**14**:6336-6340

[98] David O, Meester WJN, Bieraugel H, Schoemaker HE, Hiemstra H, van Maarseveen JH. Intramolecular staudinger ligation: A powerful ring-closure method to form medium-sized lactams. Angewandte Chemie, International Edition. 2003;**42**:4373-4375

[99] Palomo C, Aizpurua JM, Ganboa I, Oiarbide M. Asymmetric synthesis of beta-lactams through the Staudinger reaction and their use as building blocks of natural and nonnatural products. Current Medicinal Chemistry. 2004;**11**: 1837-1872

Chapter 3

Synthesis of Schiff Bases by Non-Conventional Methods

Adebimpe D. Adesina

Abstract

A Schiff base is a compound with the general structure $R_1R_2C{=}NR'$. They can be considered a subclass of imines. The term Schiff base is normally applied when these compounds are used as ligands to form coordination complexes with metal ions. Schiff bases can be synthesised from a primary aliphatic or aromatic amine and a carbonyl compound by nucleophilic addition forming a hemiaminal, followed by a dehydration to generate an imine. In other words, a Schiff base is a nitrogen analogue of a ketone or aldehyde where the carbonyl group has been replaced by azomethine or imine group. The imine group present in these compounds has been shown to be critical to their biological activities. Schiff bases have been frequently used in various fields such as medicine, pharmaceutical purposes due to their wide range of industrial applications. The unconventional methods of preparation of Schiff bases, compared with traditional methods, are more convenient, and reactions can be carried out in higher yield, shorter reaction time and milder conditions, without generation of pollution and safer to analyse.

Keywords: Schiff base, synthesis, imine, carbonyl, amine, condensation, solvent-free, natural catalysts

1. Introduction

Schiff bases are versatile C=N containing compounds possessing broad range of biological activities, and incorporation of metals in form of complexes show some degree of antibacterial, antifungal, antitumor, antiviral and anti-inflammatory properties [1]. Schiff bases are typically formed by the condensation of a primary amine and an aldehyde or ketone. Structurally, a Schiff base is a nitrogen analogue of an aldehyde or ketone in which the carbonyl group has been replaced by an imine or azomethine group. Schiff bases are some of the most widely used organic compounds. They are used as pigments and dyes, catalysts, intermediates in organic synthesis, and as polymer stabilisers [2]. They are fundamental materials for the synthesis of various ligands which can be used as chiral auxiliaries in asymmetric synthesis. Metal complexes of Schiff bases have also been used in oxidation reactions [3].

Imine or azomethine groups are present in various natural, natural-derived and non-natural compounds. The imine group present in such compounds has been shown to be critical to their biological activities (**Figure 1**) [4–6]. Schiff bases are

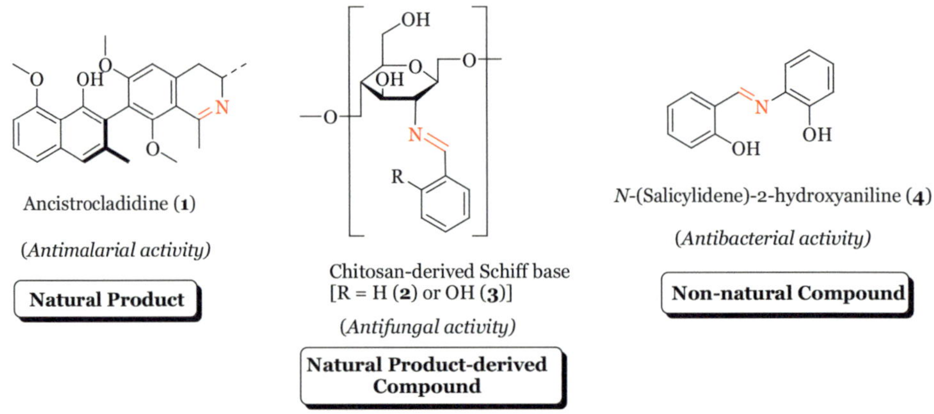

Figure 1.
Examples of bioactive Schiff bases.

represented by the general formula $R_3R_2C{=}NR_1$. The substituents R_2 and R_3 may be alkyl, aryl, heteroaryl or hydrogen while the substituent R_1 at the *N*-imino may be alkyl, aryl, heteroaryl, hydrogen or metallo (usually Si, Al, B, Sn).

The general approaches to the synthesis of Schiff bases are described in this report. Some unconventional methods of preparation and efficient practical techniques like microwave irradiation, solid-solid condensation, ultrasound irradiation, water suspension medium, infrared irradiation and the use of natural product as catalysts are discussed.

2. Preparation of Schiff bases

The most common method for the preparation of imines is the original reaction reported by Hugo Schiff in 1864 [7–9]. It is usually formed by the condensation of an aldehyde or ketone **5** with a primary amine **6** and elimination of water molecule in a Dean Stark apparatus. The formation of a Schiff base **8** is a reversible reaction and the dehydration of carbinolamine **7** generally takes place under acid or base catalysis, or upon heating. Molecular sieves are then used to completely remove water formed in the system (**Figure 2**) [10].

Aliphatic ketones react with amines to form imines more slowly than aldehydes; therefore, higher reaction temperatures and longer reaction time are required. Acid catalysts and water removal from the reaction mixture can significantly increase the

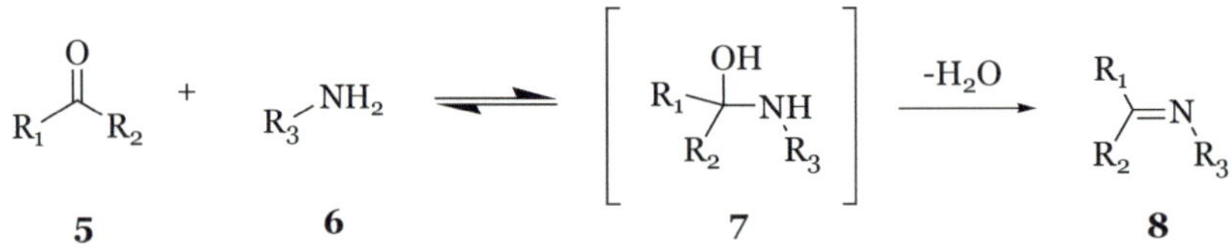

Figure 2.
Schiff reaction for the preparation of imines.

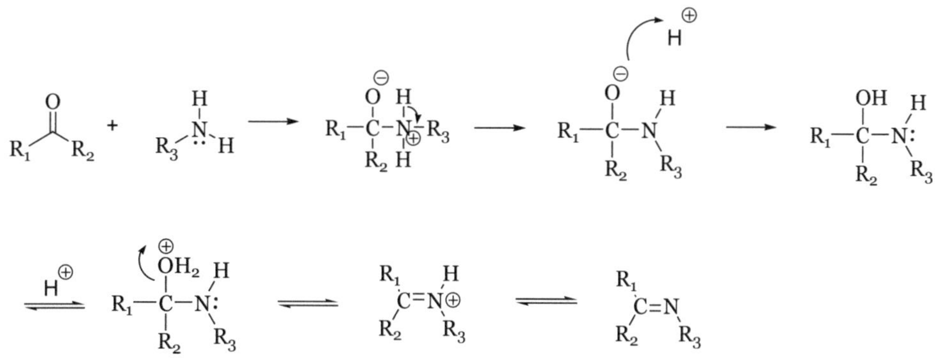

Figure 3.
Mechanism of formation of Schiff bases.

reaction yields. Aromatic ketones are less reactive than aliphatic ones and require harsh conditions to be converted into imines.

Recently, a number of innovations and new techniques for the preparation of Schiff bases have been reported including solvent-free, clay, or microwave irradiation, solid-state synthesis, molecular sieves, liquid crystals, water suspension medium, infrared and ultrasound irradiation [11–15].

The mechanism of Schiff base formation is another variation of nucleophilic addition to the carbonyl group, where the nucleophile is the amine. The amine reacts with the carbonyl to give an unstable addition compound called carbinolamine. The carbinolamine loses water by either acid or base catalysed pathways. Since the carbinolamine is an alcohol, it undergoes acid catalysed dehydration (**Figure 3**) [16].

3. Synthesis of Schiff bases from various methods

The clay catalysed synthesis of imines and enamines under solvent-free conditions using microwave irradiation has been reported by Varma et al. [11]. A synthetic procedure catalysed by montmorillonite K 10 clay was employed for the preparation of imines and enamines. An equimolar mixture of the carbonyl compound and amine was placed in an open glass container and irradiated in a microwave oven at full power for 3 minutes. The product was extracted into DCM and removal of the solvent under reduced pressure gave the imines/enamines in 75–98% yield. This approach eliminates the need for the large excess of support usually employed in solid phase reactions and reduces considerably the longer times and large quantities of aromatic solvents required in the conventional solution phase chemistry which entails the azeotropic removal of water using Dean-Stark apparatus. This method was also employed by Vass et al. [17] for the synthesis of *N*-sulfonylimines **11** (**Figure 4**). The one-pot and high-yielding protocol for the preparation of *N*-sulfonylimines from aryl aldehydes **9** and sulfonamides **10** provided a better alternative to the existing methods due to its shorter reaction time, simple reaction procedure and the formation of cleaner products that can be used for synthetic applications without further purification.

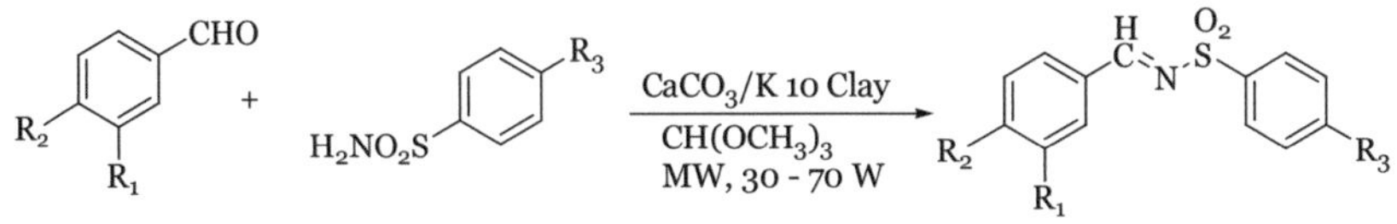

9 **10** **11**

Figure 4.
Microwave-assisted synthesis of N-sulfonylimines using calcium carbonate and K 10 clay.

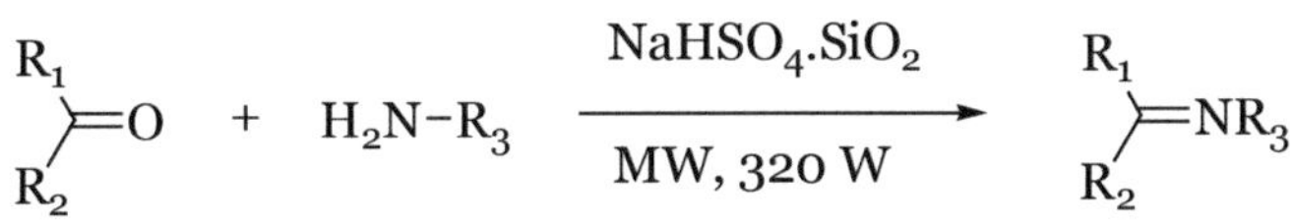

5 **6** **8**

Figure 5.
Microwave-induced synthesis of imines using $NaHSO_4.SiO_2$ catalyst.

The synthesis of imines has been carried out efficiently using silica gel supported sodium hydrogen sulfate ($NaHSO_4.SiO_2$), a non-toxic and inexpensive catalyst, as a reusable heterogeneous catalyst in solvent-free conditions under microwave irradiation [18]. Several substituted imines were prepared by this method (**Figure 5**). The $NaHSO_4.SiO_2$ catalyst can be reused by simple washing with diethyl ether after each use followed by activation in an oven at 120°C for 1 h prior to use, thus rendering the process more economical. This constitutes a green and efficient alternative to the MW assisted method described by Varma et al. [11] using K-10 clay as catalyst and the use of DCM or diethylether, which are the commonly used solvents for this reaction. The use of $NaHSO_4.SiO_2$ catalysed nucleophilic attack on the carbonyl group by the amine and served as a dehydrating agent to facilitate the removal of water in the final step. This eliminates the environmental disadvantages of using toxic drying agents such as $TiCl_4$.

Schmeyers et al. [12] reported the solid-state synthesis of various kinds of benylideneaniline derivatives **14** without passing through liquid phases. The solid-solid reactions were performed by grinding together equimolar mixture of the pure aniline **12** and aldehyde **13** in a mortar at room temperature for 2 h (**Figure 6**). Twenty preparatively useful imines were quantitatively (100% yield at 100% conversion) obtained as hydrates. The water produced in the reaction was removed at 80°C under vacuum. These solid-solid condensations, unlike (acid catalysed) imine synthesis in solution, proceed waste-free.

A mild and improved protocol for the preparation of imines by ultrasound irradiation has been developed by Guzen et al. [15]. A wide range of aromatic and heteroaromatic aldehydes were employed and all imines were obtained in excellent yields. The reaction was very efficient on a large scale, with the advantage of a very simple work-up and short reaction time (10 min). However, the process involved the use of a catalyst for activation and reaction solvent (**Figure 7**).

A new, simple, efficient, and environmentally benign method for the preparation of substituted *N*-benzylideneaniline derivatives, via condensation of several benzaldehydes and anilines, by means of infrared irradiation under solvent-free conditions

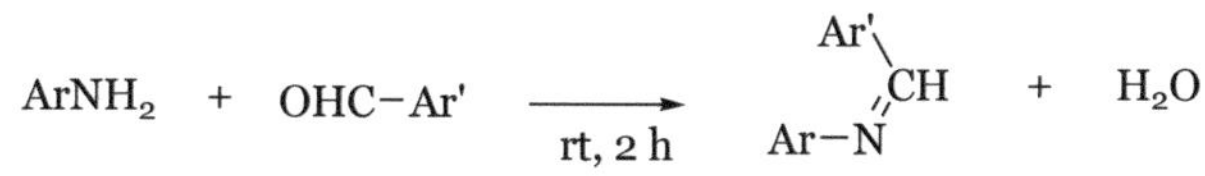

12 **13** **14**

Figure 6.
Quantitative solid-solid synthesis of azomethines.

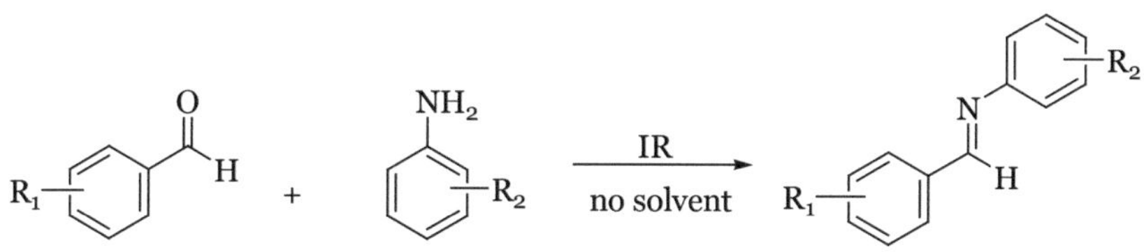

15 **16** **17**

Figure 7.
Formation of N-benzylideneanilines by IR irradiation.

has been developed [19]. The reactions proceeded with good yields and in considerably shorter times than those previously reported under classical thermal conditions with additional advantages of lower cost, ease of work-up, and the fact that activation of the reaction by an acid catalyst was unnecessary.

The synthesis of various kinds of imines by a simple and green procedure was reported by Tanaka and Shiraishi [13]. The process involved the condensation reactions of aldehydes and amines in a water suspension medium at room temperature. Water is a non-toxic, safe, and cheap medium. The use of water as solvent allowed the formation of imines without the need for catalysis, the use of a large excess of aromatic solvents, or the azeotropic removal of water. The reactions were completed in short reaction times, high-yielding and the products were isolated by filtration.

This procedure was also employed by Rao et al. [20] for the synthesis of Schiff bases via the condensation of 1,2-diaminobenzene with various substituted aromatic aldehydes in aqueous medium (**Figure 8**). The method was experimentally simple, clean, high-yielding, with reduced reaction times. The product was purified by simple filtration followed by washing with water and drying processes.

4. Synthesis of Schiff bases using natural catalysts

A green method for the synthesis of Schiff bases using natural acids found in tamarind extract and lemon juice as a catalyst has been reported (**Figure 9**) [21]. The condensation of benzaldehyde **23** with aniline **24** and with urea **26** gave benzylidene aniline **25** and benzylidene urea **27** respectively. The products which were identified by various spectroscopic techniques showed significant antimicrobial and antioxidant activities. The role of tamarind extract and lemon juice as catalysts which are important for the formation of these Schiff bases was observed in that the reaction carried out in the absence of the catalysts gave no products. The method was economical and gave high yield of product.

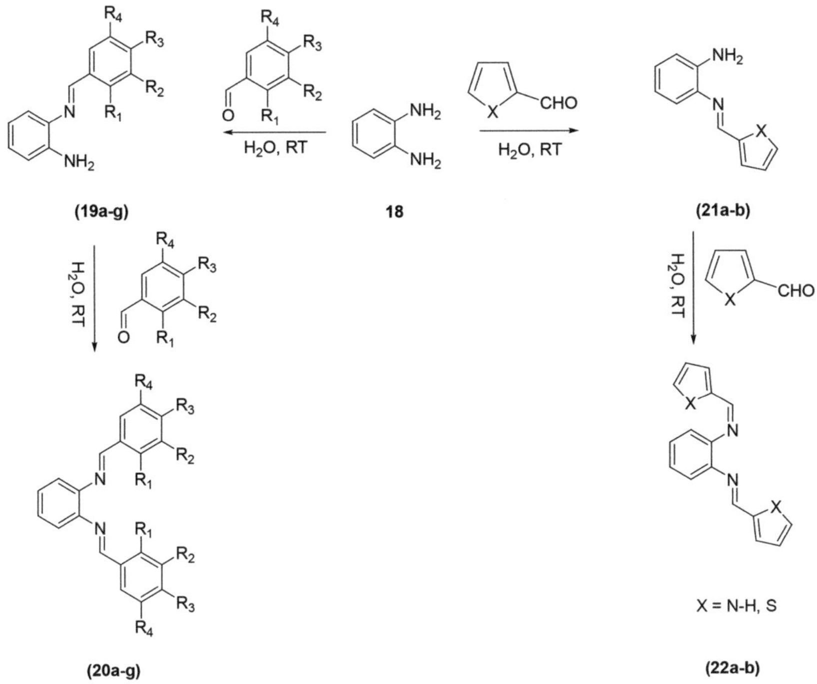

Figure 8.
Synthesis of Schiff bases in aqueous medium.

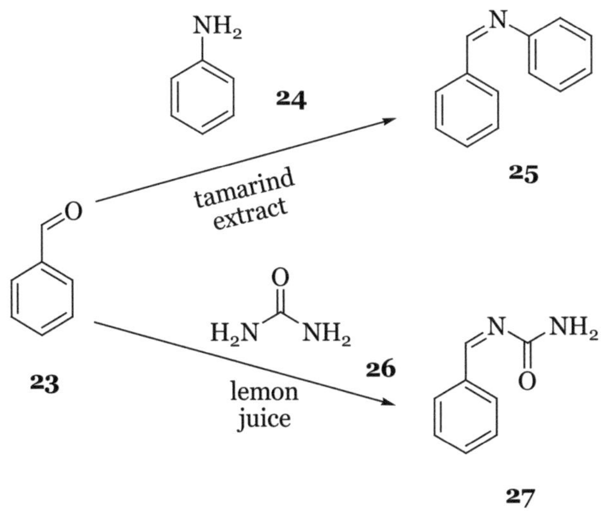

Figure 9.
Synthesis of Schiff bases catalysed by natural products.

5. Conclusion

Some non-conventional methods of preparation of Schiff bases have been extensively discussed. These methods are environment friendly, free from organic solvents and without the drawbacks of long reaction time, special apparatus and cost of dehydrating agent.

DOI: http://dx.doi.org/10.5772/intechopen.108688

Author details

Adebimpe D. Adesina
Federal University of Agriculture, Abeokuta, Ogun State, Nigeria

*Address all correspondence to: adesinaad@funaab.edu.ng

References

[1] Gupta RR, Gupta V, Kumar M. Heterocyclic Chemistry. 1st ed. Berlin Heidelberg: Springer-Verlag; 1998

[2] da Silva CM, da Silva DL, Modolo LV, Alves RB, de Resende MA, Martins CVB, et al. Schiff bases: A short review of their antimicrobial activities. Journal of Advanced Research. 2011;**2**:1-8

[3] Jarrahpour AA, Rezaei S. Synthesis of N,N'-bis (a-methylsalicylidene) 4,4′-diaminophenylmethane as a novel complexing agent. Molbank. 2006;**2006**(1):M456

[4] Bringmann G, Dreyer M, Faber JH, Dalsgaard PW, Jaroszewski JW, et al. Ancistrotanzanine C and related 5,1′- and 7,3'-coupled Naphthylisoquinoline alkaloids from Ancistrocladus tanzaniensis. Journal of Natural Products. 2004;**67**(5):743-748

[5] de Souza AO, Galetti FCS, Silva CL, Bicalho B, Parma MM, Fonseca SF, et al. Antimycobacterial and cytotoxicity activity of synthetic and natural compounds. Quimica Nova. 2007;**30**(7):1563-1566

[6] Guo Z, Xing R, Liu S, Zhong Z, Ji X, Wang L, et al. Antifungal properties of Schiff bases of chitosan, N-substituted chitosan and quartenized chitosan. Carbohydrate Research. 2007;**342**(10):1329-1332

[7] Schiff H. Mittheilungen aus dem universitatslaboratorium in Pisa: Eine neue reihe organischer basen. Justus Liebigs Annalen der Chemie. 1864;**131**:118-119

[8] Schiff H. Eine neue reihe organischer diamine. Justus Liebigs Annalen der Chemie. 1866;**140**:92-137

[9] Schiff U. Giornale di scienze naturali ed economiche. Palermo. 1867;**II**:1-59

[10] Taguchi K, Westheimer FH. Catalysis by molecular sieves in the preparation of ketimines and enamines. The Journal of Organic Chemistry. 1971;**36**:5556-5557

[11] Varma RS, Dahiya R, Kumar S. Clay catalyzed synthesis of imines and enamines under solvent-free conditions using microwave irradiation. Tetrahedron Letters. 1997;**38**:2039-2042

[12] Schmeyers J, Toda F, Boy J, Kaupp G. Quantitative solid-solid synthesis of azomethines. Journal of the Chemical Society, Perkin Transactions. 1998;**2**:989-993

[13] Tanaka K, Shiraishi R. Clean and efficient condensation reactions of aldehydes and amines in a water suspension medium. Green Chemistry. 2000;**2**:272-273

[14] Andrade CKZ, Takada SCS, Alves LM, Rodrigues JP, Suarez PAZ, Brandao RF, et al. Molecular sieves in ionic liquids as an efficient and recyclable medium for the synthesis of imines. Synlett. 2004;**12**:2135-2138

[15] Guzen KP, Guarezemini AS, Orfao ATG, Cella R, Pereira CMP, Stefani HA. Eco-friendly synthesis of imines by ultrasound irradiation. Tetrahedron Letters. 2007;**48**:1845-1848

[16] Xavier A, Srividhya N. Synthesis and study of Schiff base ligands. IOSR-JAC. 2014;7:6-15

[17] Vass A, Dudas J, Varma RS. Solvent-free synthesis of N-sulfonylimines using microwave irradiation. Tetrahedron Letters. 1999;**40**:4951-4954

[18] Gopalakrishnan M, Sureshkumar P, Kanagarajan V, Thanusu J, Govindaraju R. Silica gel supported sodium hydrogen sulfate as an efficient and reusable heterogeneous catalyst for the synthesis of imines in solvent-free conditions under microwave irradiation. Journal of Chemistry Research. 2005;**5**:299-303

[19] Vazquez MA, Landa M, Reyes L, Miranda R, Tamariz J, Delgado F. Infrared irradiation: Effective promoter in the formation of N-Benzylideneanilines in the absence of solvent. Synthetic Communications. 2004;**34**:2705-2718

[20] Koteswara Rao V, Subba Reddy S, Satheesh Krishna B, Reddi Mohan Naidu K, Naga Raju C, Ghosh SK. Synthesis of Schiff's bases in aqueous medium: A green alternative approach with effective mass yield and high reaction rates. Green Chemistry Letters and Reviews. 2010;**3**:217-223

[21] Wahab A, Haider SS, Mahmood I, Mahmood T, Sherwani SK, Kanwal S. Synthesis of Schiff bases from natural products and their remarkable antimicrobial and antioxidant activity. FUUAST Journal of Biology. 2014;**4**:27-32

Chapter 4

Schiff Bases and Their Metal Complexes: Synthesis, Structural Characteristics and Applications

Ramhari Meena, Pooja Meena, Anita Kumari, Naveen Sharma and Nighat Fahmi

Abstract

The development of Schiff base was a major step forward in the area of coordination chemistry. Schiff bases, a class of organic compounds, carry the imine or azomethine (>C=N–) functional group. Schiff bases played an influencing role in the development of coordination chemistry and were a key point in the development of inorganic, bioinorganic chemistry and optical materials. Schiff bases, widely used in inorganic, organic, and analytical chemistry, account for a significant portion of the more commonly employed classes of organic molecules. The ability of Schiff base ligands to form stable metal complexes with a wide range of transition and other metal ions makes them extremely useful. Condensation of a primary amine with an aldehyde or ketone yields a Schiff bases. In this chapter, we focused on introducing Schiff bases, classified them and their metal complexes, and discussed several synthesis methods, including conventional and green approaches. This chapter also elaborated on the industries' applications, such as the food industry, agrochemical industry, dye industry, analytical chemistry, catalysis, energy storage, environmental, chemo-sensing, bio-sensing, and biomedical applications of novel Schiff bases and their metal complexes.

Keywords: Schiff base, metal complexes, synthesis, structure, industrial and biomedical applications

1. Introduction

Hugo Schiff, a German chemist, initially reported Schiff base in 1864 [1]. Schiff bases are organic molecules formed via condensation reaction of carbonyl compounds and primary amines [2]. The typical structure can be expressed as R'-CR = N-R", where R, R' and R" might vary. R and R' could be alkyl, aryl and heterocyclic structures with various substituents. The carbonyl group may be a constituent of an aldehyde or a ketone (>C=O)). Because Schiff bases contain an azomethine (>C=N-)) group, they are also known as azomethine or imine. The Schiff bases derived from aldehydes and ketones are known as aldimines and ketimines, respectively. A general

Schiff base condensation reaction involving the amine and the carbonyl functional group could be represented as follows (**Figure 1**).

When synthesizing Schiff bases, nucleophilic amines are used to attack electrophilic carbonyl compounds via a nucleophilic addition process, forming a hemiaminal group, and then the hemi-aminal group is dehydrated to generate imine compounds. In the first phase of the reaction, the amine reacts with the aldehyde or ketone to generate the unstable addition product carbinolamine. Carbinolamine undergoes acid- or base-catalyzed dehydration. Considering that carbinolamine is an alcohol, it undergoes a dehydration reaction when subjected to an acid catalyst (**Figure 2**). Reversible acid or base catalysis or heating often happens during the production of a Schiff base from aldehydes or ketones. When the product is isolated or water is evaporated, or both, the formation is pushed to completion. Hydrolysis of various Schiff bases by an aqueous acid or base yields the corresponding aldehydes, ketones, and amines.

Schiff bases with aryl substituents are significantly more stable and easier to synthesize, but those with alkyl substituents are relatively unstable [3, 4]. Aliphatic aldehyde Schiff bases are highly unstable and easily polymerizable, but aromatic aldehyde Schiff bases with efficient conjugation are more stable. Because aldehydes have less stearic hindrance than ketones, thus react faster. Extra carbon in ketones makes them less electrophilic than aldehydes. Schiff bases have received a great deal of attention due to their simplicity of synthesis, availability, and electronic characteristics. There is great interest in developing a wide range of applications in organic [5], inorganic [6, 7], coordination [8–10], bioinorganic [11, 12] and environmental chemistry [13–16]. Schiff base derivatives have been utilized in medical, pharmaceutical, metal refining, metallurgy, catalysis, food, sensing, filtration, environmental, photography and diagnostic applications.

The discovery of Schiff base was a major step forward for the discipline of coordination chemistry. When combined with a variety of transition metal ions, Schiff base ligands can generate stable metal complexes with a wide range of applications. The Schiff base has remarkable chelating characteristics. The presence of hydroxyl and thio-groups in azomethine groups may result in forming a penta or hexa ring structure with metal ions. Bidentate, tridentate, tetradentate, and polydentate Schiff bases are also possible. Because of the availability of a lone pair on the nitrogen atom, the Schiff base forms complexes with many metals. This lone pair aids in creating monodentate complexes, while adding other groups, such as OH and SH, may result in the formation of bidentate chelates. The azomethine nitrogen atom's lone pair of electrons and sp^2 hybridized. As a result, it has substantial biological and chemical significance. Because there are more donor atoms in heterocyclic rings containing Schiff bases, they play a larger role in coordination chemistry [17–19].

Schiff bases show biological activities like nematicidal [20], insecticidal [21], antibacterial [22], antifungal [23], antileukaemia [24], anti-inflammatory [25], anti-HIV activity [26], antimycobacterial activity [27], antioxidant [28],

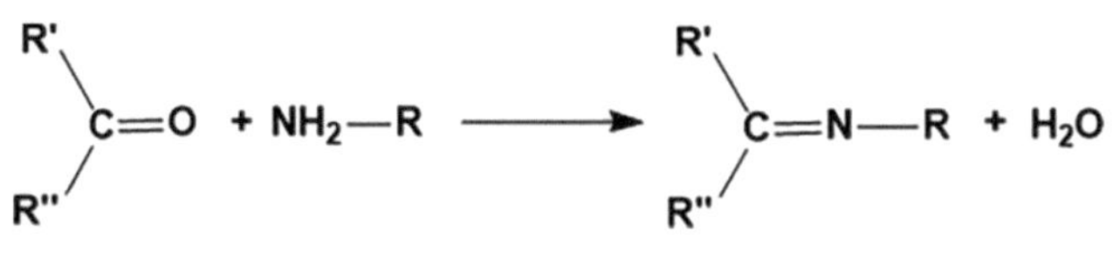

Figure 1.
General scheme of formation of Schiff bases.

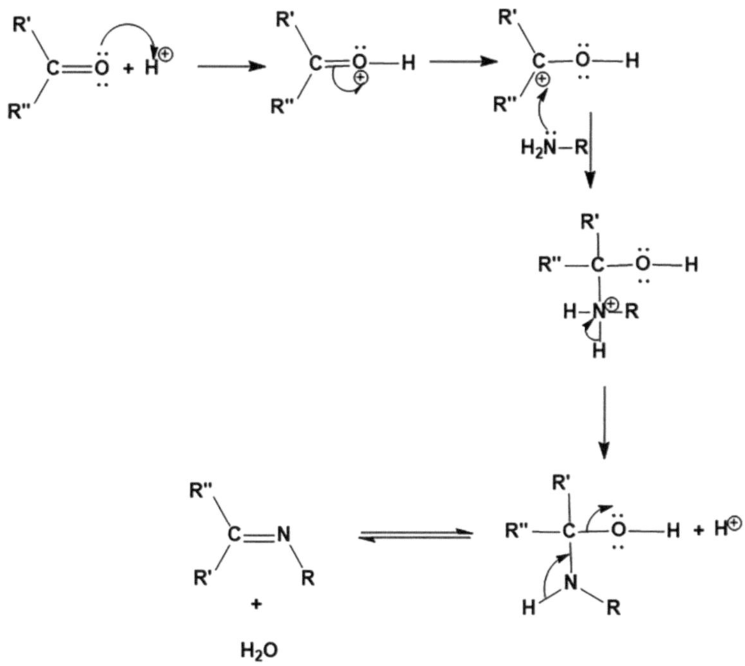

Figure 2.
Mechanistic explanation of the formation of Schiff base.

anticancer [29], and plant growth regulatory activity [30] among others. Besides biological applications, Schiff base and their metal complexes have enormous applications in analytical chemistry [31], dye industry [32], and corrosion inhibitors [33]. Schiff base has attracted the attention of many experimental and theoretical researchers due to their particular photo-luminescence [34] in the visible range and at room temperature and so applied in many realms such as microelectronics [35], optoelectronics [36] and biological sensors [37]. Schiff base shows excellent catalytic activity in various reactions such as polymerization reaction and reduction of thionyl chloride, reduction reaction of ketones, oxidation of organic compounds, aldol reaction, epoxidation of alkenes, hydrosilylation of ketones, Henry reaction, synthesis of bis (indolyl) methanes and Diels Alder reaction [38–40]. Synthetic chemists have used Schiff bases and related complexes for a wide variety of processes, including the oxidation of alkenes and the catalytic transformation of hydrocarbons into useful oxygenated derivatives such alcohols, aldehydes, and epoxides. A further area of intense curiosity is the catalysis of alkene oxidation by soluble transition metal complexes. Schiff bases have a wide range of donor sites, resulting in significant transition metal complexes. Schiff base metal complexes have been used in biological applications, resulting in important recent improvements in a variety of chemistry areas. Because of their unique treatment method, metal-containing antibacterial compounds appear to be promising candidates for brand-new antibiotic medications that restrict the growth of bacterial strains. The characteristics of Schiff base metal complexes vary depending on the ligands and the transition metal ion. Schiff bases have attracted a lot of interest because of their various chemical and physical properties, as well as their ease of production.

2. Synthetic methods of Schiff bases

Schiff base ligands, a class of molecules having imine groups, have grown in popularity due to their physiological and pharmacological properties. They are a fascinating class of chelating agents capable of coordinating metal ions in a complex, which is used to imitate biological processes. Many studies have been conducted on synthesizing Schiff bases [41–43]. Schiff bases have been prepared using conventional and green synthetic methods (**Figure 3**).

Heat is required in many condensation processes, and traditional reaction conditions often involve heating the reactants in a metal, oil, or sand bath for hours or even days. The conventional procedure involves refluxing or stirring different aldehydes or ketones with various types of primary amines. Green chemistry refers to the tools and procedures that provide considerable environmental and financial benefits over conventional synthetic approaches. It depicts that the current in green chemistry has triggered a new demand for organic synthesis in which distinct reaction environments must be located, reducing the usage of harmful organic solvents or toxic chemicals. Green approaches must improve selectivity, reduce reaction time, and simplify product isolation over conventional methods. Microwave-assisted synthesis of Schiff bases has been carried out without solvent or low-solvent conditions and reduces reaction time significantly, improves conversion, and sometimes increases selectivity. Since the development of solvent less microwave synthesis of Schiff bases, it has become the most well-known and simple technique for these reactions and is used in various applications. Many researchers reported using the microwave-assisted synthesis of various types of Schiff bases and their derivatives.

The grindstone technique reaction creates local heat by grinding substrate crystals and reagent with a mortar and pestle. Grinding starts reactions by transmitting a

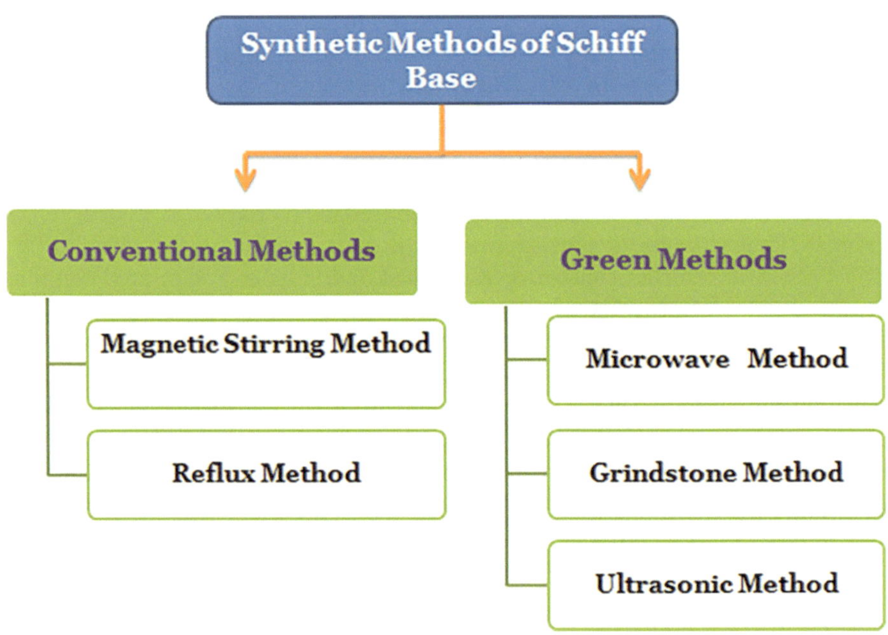

Figure 3.
Synthetic methods of Schiff bases.

relatively small quantity of energy through friction. In some circumstances, a mixture and reagents form a glassy substance. Such reactions are simple to handle, eliminate pollutants, are relatively cheaper to operate, and may be considered more economical and environmentally friendly in chemistry [18]. Because molecules in a crystal are organized tightly and regularly, solid-state reactions are more efficient and selective than solution reactions [19]. The synthesis of amino acid Schiff bases (3) in water by the reaction of variously substituted aromatic aldehydes/heterocyclic aldehydes (2) and dl -alanine amino acid (1) stirring at room temperature (method A) using grindstone chemistry (method B), microwave irradiation (method C), and conventional heating (method D). Studies comparing the times required to synthesize four distinct Schiff bases under four different conditions found that while method B (grindstone) had a better product yield, it also took significantly longer than procedures A, C, and D., i.e., Product yields ranged from 72 to 78% after being subjected to grinding for 30–35 minutes followed by leaving the reaction mixture overnight (8 h). Product yields of 69–73% were achieved in 40–45 minutes when the reactants were stirred in water at room temperature (15–20°C). Microwave irradiation (method C) and ordinary heating (method D) were also used to synthesize compounds. In 5–6 minutes, yields of 70–72% were achieved under microwave irradiation, which is much quicker than grinding (method B). The highest yield and shortest reaction time were both achieved by using Method C (**Figure 4**).

Sonication is the use of high power to excite particles for various purposes. Ultrasonics with frequencies greater than 20 kHz is normally employed in a process known as ultrasonication [45]. In the laboratory, it is typically used with an ultrasonic bath or probe; this apparatus is known as a sonicator. A new approach for

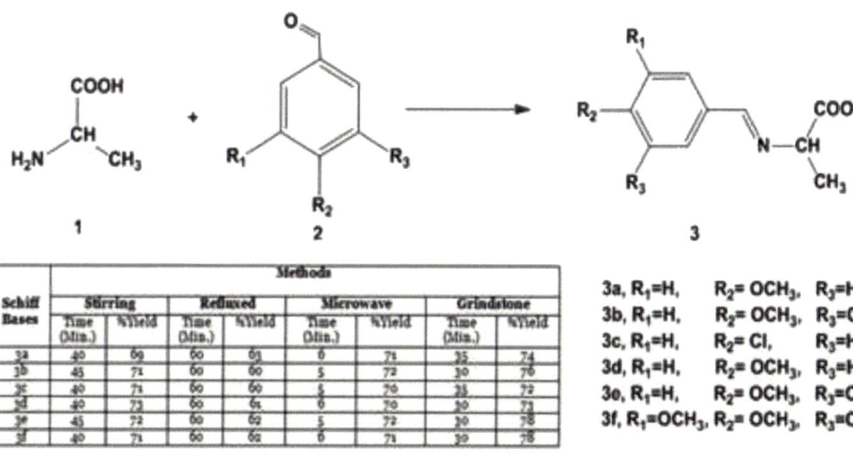

Schiff Bases	Methods							
	Stirring		Refluxed		Microwave		Grindstone	
	Time (Min.)	%Yield	Time (Min.)	%Yield	Time (Min.)	%Yield	Time (Min.)	%Yield
3a	40	69	60	63	6	71	35	74
3b	45	71	60	60	5	72	30	76
3c	40	71	60	60	5	70	35	72
3d	40	73	60	61	6	70	30	73
3e	45	72	60	62	5	72	30	78
3f	40	71	60	62	6	71	30	78

3a, R_1=H, R_2= OCH_3, R_3=H
3b, R_1=H, R_2= OCH_3, R_3=OH
3c, R_1=H, R_2= Cl, R_3=H
3d, R_1=H, R_2= OCH_3, R_3=H
3e, R_1=H, R_2= OCH_3, R_3=OCH_3
3f, R_1=OCH_3, R_2= OCH_3, R_3=OCH_3

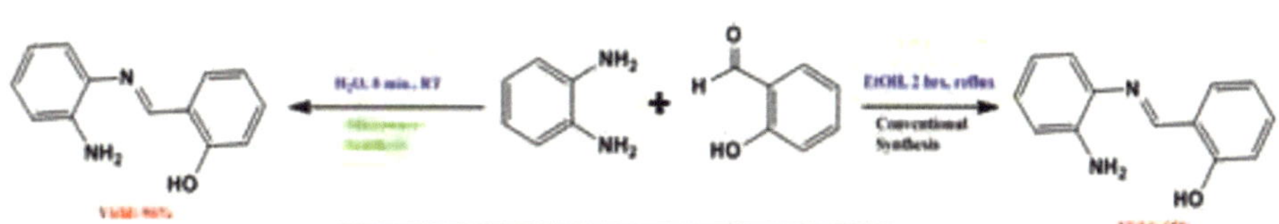

Methods	Time	%Yield
Conventional Method	2h	65
Microwave Assisted Method	8min.	96

Figure 4.
Synthesis of Schiff bases 3a–3f with different methods and comparisons [44].

synthesizing Schiff bases under catalyst-free ultrasonic irradiation conditions has been discovered, yielding 92% compared to the current method's 84% yield [46]. They discovered an efficient and environmentally friendly method for Schiff base synthesis in an aqueous solution using ultrasonic irradiation conditions and thus no catalyst. Several research publications revealed how different green synthesis techniques could cause a specific condensation process to occur preferentially. When opposed to solution-based synthesis, mechanochemical synthesis has the advantage of ease of set-up and high yields [47]. Most Schiff bases have reported more excellent ligation with metal ions because of a lone pair of electrons in these compounds. The number of transition and other metal ions complexes were synthesized by using novel Schiff base ligands [17].

3. Classification of Schiff bases and their metal complexes

By coordinating the d-block metal ion with the electron-donating ligand atom, a complex is created that modifies the metal's steric and electrical properties. By doing so, the metal ion's reactivity is stabilized and regulated, which is especially helpful for ions at higher oxidation states where they are less stable. Auxiliary ligands, or Schiff bases, are compounds that modify the structure and reactivity of a transition metal ion inside a complex. On the other hand, they do not incur any irreversible modifications, unlike reactive ligands. Atoms like nitrogen, sulfur, or oxygen can act as donors in the coordination process.

3.1 Denticity of Schiff base ligands

Schiff bases are classified into bidentate, tridentate, tetradentate, and polydentate ligands, which can form extremely stable complexes with transition metal ions (**Figure 5**). Assume they have different functional groups such as -OH, $-NH_2$, or -SH; the resulting Schiff bases can act as mixed-donor ligands in bi-, tri-, tetra-, and higher

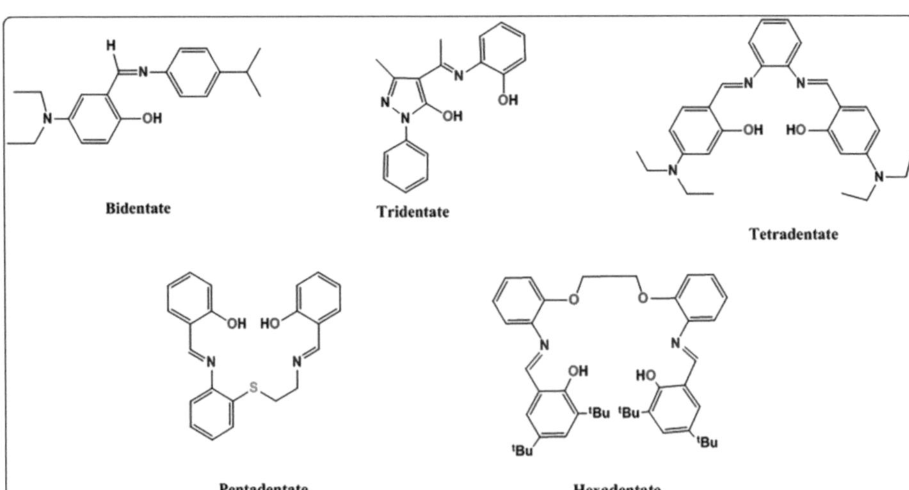

Figure 5.
Classification of Schiff bases into bidentate, tridentate, tetradentate, and polydentate ligands.

coordination modes [48–51]. Multivalent Schiff base ligands easily form complexes with bidentate, tridentate, and tetra- or polydentate metal ions at different oxidation states. Donor atoms (N, O, S) can be found in bidentate ligand (NN or ON), tridentate ligands (NNN, ONO, NNS or ONS), and tetradentate ligands (ONNO, NNNN, NSNO).

3.2 Symmetrical and asymmetrical Schiff bases

The Schiff bases' great affinity for chelation towards inner and non-inner transition metal ions is employed to produce stable complexes. Unsymmetrical ligands do not have a rotation or mirror axis of symmetry and bind to a metal ion with two groups, and symmetrical ligands have [52, 53]. It has been argued that unsymmetrical Schiff base ligands are superior to their symmetrical counterparts due to their ability to more accurately predict the geometry of metal ion binding sites in biological systems that contain metal ions and their ability to more easily combine natural and synthetic structural components (**Figure 6**).

3.3 Homoleptic and heteroleptic Schiff base metal complexes

The primary distinction between homoleptic [54] and heteroleptic [55] complexes is that homoleptic complexes have identical ligands linked to a metal centre.

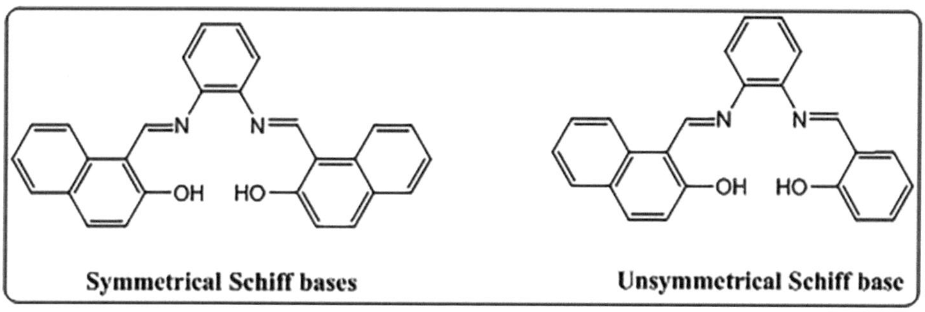

Figure 6.
Symmetrical and asymmetrical salphen Schiff bases.

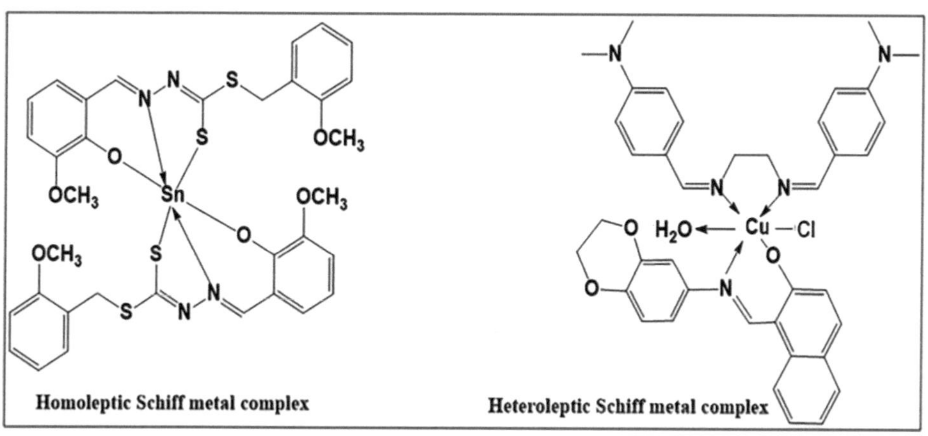

Figure 7.
Homoleptic and heteroleptic Schiff base metal complexes.

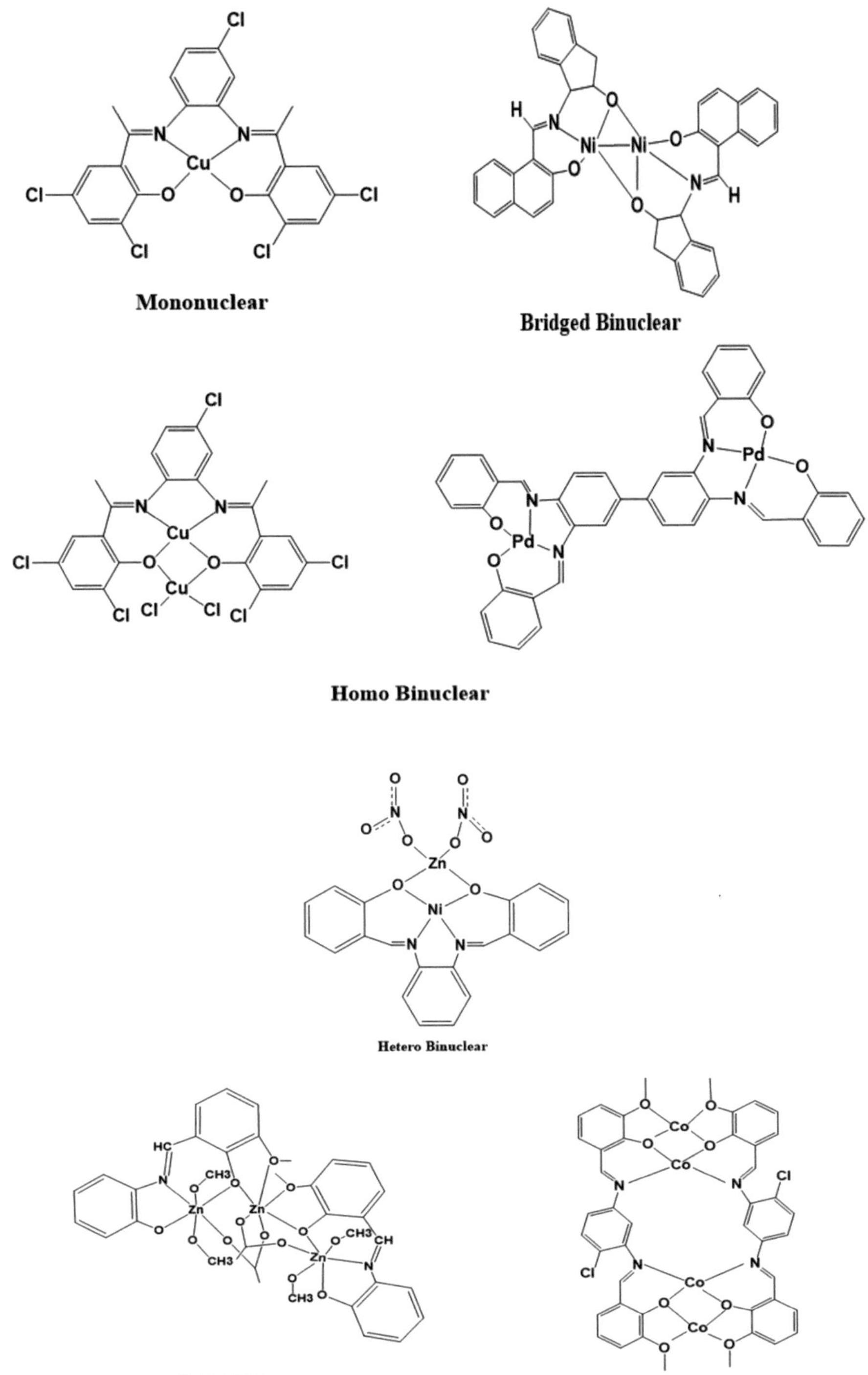

Figure 8.
Mononuclear and polynuclear Schiff base metal complexes [56–61].

DOI: http://dx.doi.org/10.5772/intechopen.108396

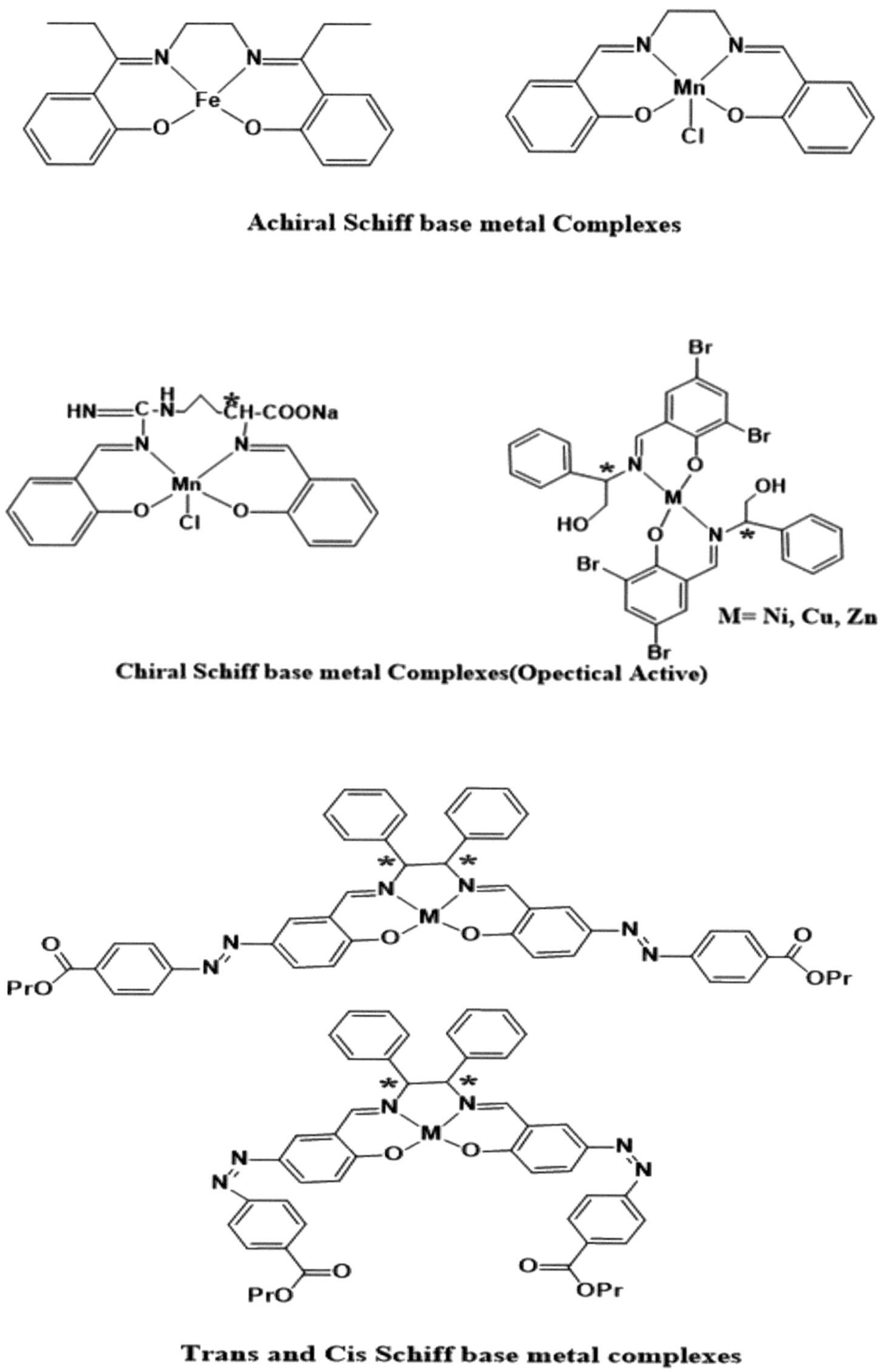

Figure 9.
Achiral, chiral, trans and cis Schiff base metal complexes [62–65].

In contrast, heteroleptic complexes have at least one distinct ligand coupled to the complex's metal centre (**Figure 7**).

3.4 Mononuclear and polynuclear Schiff base metal complexes

A single metal atom or ion is contained within the most basic type of Schiff base metal complex, and it is surrounded by monodentate, bidentate, tridentate, and polydentate ligands. Polynuclear Schiff base metal complexes are attributed to the presence of two or more atoms of metal, or ions, co-ordinated within a single coordination sphere. The two atoms may be linked together by direct metal-metal bonds, bridging ligands, or all of these things. As versatile ligands, Schiff bases form various polynuclear metal complexes such as homonuclear and heteronuclear. These flexible ligands have the ability to act as monodentate, bidentate, or polydentate, and they can be engineered to produce mononuclear, dinuclear, or polynuclear metal-organic frameworks. It is possible to change the nuclearity of Schiff base complexes; for example, it is possible to synthesize either mono- or dinuclear complexes using nearly identical ligands and synthetic processes for both types of complexes (**Figure 8**).

3.5 Achiral and chiral Schiff base metal complexes

A chiral Schiff base metal complex is not superimposable with its own mirror image because the two structures are not identical in all respects. The mirror image of an achiral Schiff base metal complex is identical to the complex itself and can be superimposed on it. The phenomena of optical activity have traditionally been defined in terms of asymmetry and dissymmetry; however, the term chirality has recently superseded these earlier classifications. Chiral entities exist as two species with the same chemical constitution. The only way they are distinguishable from one another is that they have the opposite configuration of an object and the mirror image of that thing. Chemical compounds can be said to be stereoisomers if their chemical constitutions are same but their spatial arrangements of their atoms are different. Chiral refers to the property of molecules that prevents them from being brought into coincidence with their mirror copies by the use of stiff motions (**Figure 9**).

4. Application of Schiff bases and its metal complexes

Schiff bases and their metal complexes find widespread use in various industries and applications, including the food industry, the agrochemical industry, the dye industry, analytical chemistry, catalysis, energy storage, environmental, chemosensing, bio-sensing, nanotechnology, and biomedical applications.

4.1 Catalysis

Catalytic activity is enhanced in both homogeneous and heterogeneous reactions by Schiff base metal complexes. The ligands, coordination sites, and metal ions employed in a given compound determine its activity. Many different reactions, such as polymerization, ring-opening polymerization, oxidation, epoxidation, allylic alkylation, reduction of ketones, hydrazination of acetophenones, the Michael addition reaction, the decomposition of hydrogen peroxide, the annulation reaction, the Heck reaction, the carbonylation reaction, and the Diels-Alder reaction, have been used to critically evaluate the catalytic activity of metal complexes. There is significant potential for Schiff base ligands to be used as metal complexes in catalysis due to their simple synthesis method and heat stability. The catalytic activity of Schiff base complexes differed greatly depending on the structure and kind of ligands used [66, 67].

4.2 Dye industry

The dyeing technique employs a wide range of Schiff bases and complexes, many of which have been synthesized, investigated, and employed as mordants [68]. As a dye, transition metal complexes such as iron (III), nickel (II), cobalt (II), and copper (II) complexes, among others, have been prepared from a variety of Schiff bases and employed to produce a variety of transition metal complexes. Aldehyde groups that include azo dyestuff are known to synthesized many azomethine linkages that contain azo dyes due to condensation with primer amines. The textile industry utilizes these dyestuffs to color a variety of materials. Outside the textile sector, the field of photochemistry places a significant emphasis on using azo dyes that include the amine group. The Schiff base on fluorene showed desirable properties including sensitivity to pH, as well as heat and color stability. For making a water-based ink, it showed promise as a functional pigment material [69].

4.3 Food industry

Various research groups have recently concentrated on producing natural novel and active materials for food packaging applications. Because of their antibacterial action, chitosan-derived Schiff base films developed may not only boost the safety of such foods and hence lengthen their shelf life, but also provide a flavor that is well-accepted by the consumer. Schiff's base (SB) modified zirconium dioxide reinforced PLA bio-composite film serves as an alternate packing material to replace single waste synthetic manufactured materials that pollute the environment. For active packaging applications, Schiff base (SB) modified polylactic acid (PLA) film can provide improved barrier and antifungal qualities [70–72].

4.4 Agrochemical industry

Metal complexes with diverse Schiff base ligands have attracted the attention of chemists in recent years due to their agricultural applications, such as pesticidal, nematicidal, and insecticidal. Unsymmetrical Schiff bases glyoxal salicylaldehyde succinic acid dihydrazide and its Ni(II), Co(II), Zn(II), and Cu(II) complexes have been synthesized and studied; at greater concentrations, they display considerable insecticidal action [73]. H2L [2, 2′-[(1E, 2E)-ethane-1,2- diylidenedi (E) azanylylidene] dibenzenethiol] and its new Zn(II), Ni(II) metal complexes have been employed as insect repellent agents [74]. Coumarin-based Schiff base and its earth metal complexes [75] have been used to treat pests (Tribolium castaneu) and worms (Meloidogyne incognita).

4.5 Analytical applications

Schiff bases have been used as analytical probes or reagents by researchers. These are used to analyze primary amines, carbonyl compounds, and functional groups. In complexes, azomethine bonds are formed through complex formation reactions or changes in their spectroscopic properties caused by pH and solvent variations (pH of solvent polarity indicators). Schiff bases are a great carrier for the selective and efficient extraction of certain metal ions. They are well-known for their effective chelating capabilities. Schiff bases extract metal ions, essential in regulating heavy metal pollution. N, N -bis(3-methylsalicylidene)-ortho-phenylene diamine, Schiff base used in spectrophotometric detection of nickel. The approach has been used successfully to quantify trace quantities of nickel in natural food samples [76]. Schiff bases are renowned for their ability to form complexes and serve as good chelating ligands. They have been widely employed as analytical reagents due to their ligation property. Schiff bases made of salicylaldehyde are employed in gravimetric and spectrophotometric analyses. In addition, the same reagent was recently employed for the spectrophotometric detection of Ni (II) at a trace level. $Cu2^{+}$ ions have been detected using the fluorescent 4-(1-phenyl-1-methylcyclobutane3-yl)-2-(2-hydroxy-5-bromobenzylidene) aminothiazole Schiff base. This chemical sensor operates in the visible region, has a wide dynamic operating range, and may be used over a wide pH range [77].

4.6 Energy storage

There is a resurgence of interest in the search for effective, clean, and sustainable energy sources (like wind and solar) as well as cutting-edge energy conversion and storage technologies as a result of the rapid growth of the world economy, the depletion of fossil fuels, and rising environmental pollution. Energy storage technologies are more important in our lives since the sun does not shine at night and the wind does not blow all the time. Currently, there is a lot of interest in electrical energy storage technologies including batteries and electrochemical capacitors (supercapacitors). Recent research has shown that organic oligomeric Schiff bases and electroactive polymeric (linear or hyperbranched) Schiff bases perform satisfactorily as negative electrodes (anodes) in sodium-ion batteries [78]. Lithium-ion batteries have also made use of nitrogen-rich carbon nanosheets produced by the Schiff base reaction in a molten salt solution as anode materials [79]. The linear polymeric Schiff bases developed by Armand et al. [80] as a consequence of the condensation of aromatic dialdehydes with aliphatic and aromatic diamines performed well as anodes for sodium-ion batteries. Polymeric Schiff bases are also produced by combining terephthalic-aldehyde, phenylenediamine, and polyether amine blocks, resulting in polymers with high adhesive qualities that can be used as redox-active binders for sodium-ion anodes. Similarly, Zhang et al. [81] developed another ImCOF (Imine bonds containing *covalent organic frameworks)* that performed again as an anode material for lithium-ion batteries derived from 2,4,6-triaminopyrimidine and terephthalaldehyde.

4.7 Environmental applications

Most firms worldwide need copper, silver, lead cadmium, aluminum and cobalt. These metals can be present in nearly all dairy products. Their widespread prevalence in industrial processes, chronic metal contamination from occupational contact, and health risks associated with these metals necessitate their identification and control in biological and dietary samples. Metals are used in electroplating, alloy production, and battery manufacturing. As a result, excessive metal concentrations have been discovered in diverse water sources, soil, and plants. Products such as cigarettes, beers, oils, and supplements necessitate metals monitoring and quality control [82].

Metal corrosion has a tremendous impact on the national economy and critical safety and pollution issues. Although many inhibitors have good inhibitory properties, they are insufficient for environmental protection and sustainable development initiatives for various reasons (such as difficulty in degradation, toxicity or high-temperature resistance). Stable, efficient, and ecologically friendly inhibitors are the inhibitors of the future. Many inhibitors, including imidazolines, Mannich bases, and Schiff bases, contain heteroatoms (N, S, O) or chemical interactions with electrons (p bonds). N, O, and S heteroatoms, as well as unsaturated >C=N- bonds, can create strong and durable corrosion-inhibiting adsorption films on metal surfaces, demonstrating outstanding inhibitory effects. At the same time, Schiff base compounds are attractive to researchers due to their inexpensive cost, ease of synthesis and purification, strong water solubility, and low toxicity [83].

4.8 Chemo-sensing applications

Schiff base-based fluorescent probes have recently been invented for detecting and monitoring numerous hazardous analytes in biological systems. Schiff base compounds with nitrogen-oxygen-rich coordination as a receptor site provide a stable platform for fluorescence sensing with significant, visible color shifts. Detecting metal ions with diverse mechanisms in an accurate sample using Schiff base-based sensors is appealing currently. In the recent decade, Schiff base probes based on fluorescence live-cell imaging have been used to detect metal ions such as $Co2^{+}$, $Cu2^{+}$, $Zn2^{+}$, $Hg2^{+}$, Ag^{+}, $Al3^{+}$, and ClO^{-} ions [84–86].

4.9 Bio-sensing applications

Within cells, Schiff base compounds have been used as biosensors for H_2O_2, glucose, and Oncomarker CA-125 [87]. Evaluation of the sensitivity and specificity of *the gold Schiff base complex-doped sol gel nano optical sensor* for the detection of CA-125 in ovarian cancer patient samples was performed and compared to results obtained from samples taken from healthy women serving as a control group [88]. Sheta M. Sheta et al. created an ultrasensitive method of detecting human creatinine using a cerium(III)-isatin Schiff base complex as an optical sensor [89].

4.10 Biomedical applications

Schiff bases and their metal complexes have numerous applications in various biomedical pharmaceuticals such as antimicrobial, anti-malarial, anticancer, antiviral, anti-inflammatory, antioxidant, anticonvulsant, anti-anthelmintic, bioprinting, tissue regenerating, enzyme inhibition and drug delivery. In biological systems, the azomethine nitrogen of Schiff bases serves as a binding site for metal ions to attach to diverse biomolecules such as proteins and amino acids for anti-germ activity. Our bodies' Schiff bases catalyzed many metabolic events in the form of enzymes that are active against certain bacteria. Several studies have been conducted to improve the bio-functions of Schiff bases and their metal complexes. Schiff bases can fight cancer, fungus, germs, ulcers, and viruses, depending on which transition metal ions they contain [90–92].

5. Conclusion

Schiff bases and their metal complexes formed are the essential components of coordination and bioinorganic chemistry. The food industry, the agrochemical industry, the dye industry, analytical chemistry, catalysis, energy storage, the environmental field, chemo-sensing, bio-sensing, and the biomedical sector are some of the domains in which these have been used. This chapter discusses the conventional synthesis and environmentally friendly synthesis, as well as the categorization of Schiff bases and their metal complexes. Research into its applications in other fields of interest, such as catalysis, metal ion sensing, and cell imaging, makes its study relevant and worthy of being pursued.

Author details

Ramhari Meena*, Pooja Meena, Anita Kumari, Naveen Sharma and Nighat Fahmi
Department of Chemistry, University of Rajasthan, Jaipur, India

*Address all correspondence to: ramharichem@gmail.com

References

[1] Schiff H. Mittheilungen aus dem Universitats-laboratorium in Pisa (a report from the University Laboratory in Pisa). Justus Liebigs Annalen der Chemie. 1864;**131**:118-119

[2] Jain S, Rana M, Sultana R, Mehandi R, Rahisuddin. Schiff Base metal complexes as antimicrobial and anticancer agents. Polycyclic Aromatic Compounds. 2022;**5**:1-56. DOI: 10.1080/10406638.2022.2117210

[3] Hussain Z, Yousif E, Ahmed A, Altaie A. Synthesis and characterization of Schiff's bases of sulfamethoxazole. Organic and Medicinal Chemistry Letters. 2014;**4**(1):1-4

[4] Da Silva CM, da Silva DL, Modolo LV, Alves RB, de Resende MA, Martins CV, et al. Schiff bases: A short review of their antimicrobial activities. Journal of Advanced Research. 2011;**2**(1):1-8. DOI: 10.1016/j.jare.2010.05.004

[5] Jain A, De S, Barman P. Microwave-assisted synthesis and notable applications of Schiff-base and metal complexes: A comparative study. Research on Chemical Intermediates. 2022;**48**(5):2199-2251. DOI: 10.1007/s11164-022-04708-7

[6] Ghobakhloo F, Azarifar D, Mohammadi M, Keypour H, Zeynali H. Copper (II) schiff-base complex modified UiO-66-NH2 (Zr) metal–organic framework catalysts for Knoevenagel condensation–Michael addition–Cyclization reactions. Inorganic Chemistry. 2022;**61**(12):4825-4841. DOI: 10.1021/acs.inorgchem.1c03284

[7] Jasim SA, Riadi Y, Majdi HS, Altimari US. Nanomagnetic macrocyclic Schiff-base–Mn (ii) complex: An efficient heterogeneous catalyst for click approach synthesis of novel β-substitued-1, 2, 3-triazoles. RSC Advances. 2022;**12**(28):17905-17918. DOI: 10.1039/d2ra02587f

[8] Kargar H, Fallah-Mehrjardi M, Behjatmanesh-Ardakani R, Munawar KS, Ashfaq M, Tahir MN. Synthesis, spectral characterization, SC-XRD, HSA, DFT and catalytic activity of a dioxidomolybdenum complex with aminosalicyl-hydrazone Schiff base ligand: An experimental and theoretical approach. Polyhedron. 2021;**208**:115428. DOI: 10.1016/j.poly.2021.115428

[9] Mohamed GG, Omar MM, Moustafa BS, AbdEl-Halim HF, Farag NA. Spectroscopic investigation, thermal, molecular structure, antimicrobial and anticancer activity with modelling studies of some metal complexes derived from isatin Schiff base ligand. Inorganic Chemistry Communications. 2022;**141**:109606. DOI: 10.1016/j.inoche.2022.109606

[10] Aragón-Muriel A, Reyes-Márquez V, Cañavera-Buelvas F, Parra-Unda JR, Cuenú-Cabezas F, Polo-Cerón D, et al. Pincer complexes derived from tridentate Schiff bases for their use as antimicrobial Metallopharmaceuticals. Inorganics. 2022;**10**(9):134. DOI: 10.3390/inorganics10090134

[11] Shahraki S. Schiff base compounds as artificial metalloenzymes. Colloids and Surfaces B: Biointerfaces. 2022;**218**:112727. DOI: 10.1016/j.colsurfb.2022.112727

[12] Rangaswamy J, Ankali KN, Naik N, Nuthan BR, Satish S. The Mn (II), Co (II), Ni (II) and Cu (II) complexes of (Z)-N'((1H-indol-3-yl) methylene)

nicotinohydrazide Schiff base: Synthesis, characterization and biological evaluation. Journal of the Iranian Chemical Society. 2022;**19**:3993-4004. DOI: 10.1007/s13738-022-02580-1

[13] Farag AA, Toghan A, Mostafa MS, Lan C, Ge G. Environmental remediation through catalytic inhibition of steel corrosion by Schiff's bases: Electrochemical and biological aspects. Catalysts. 2022;**12**(8):838. DOI: 10.3390/catal12080838

[14] Liu H, Ding S, Lu Q, Jian Y, Wei G, Yuan Z. A versatile Schiff Base Chemosensor for the determination of trace Co2+, Ni2+, Cu2+, and Zn2+ in the water and its bioimaging applications. ACS Omega. 2022;7(9):7585-7594. DOI: 10.1021/acsomega.1c05960

[15] Paul A, Silva TA, Soliman MM, Karačić J, Šljukić B, Alegria EC, et al. Benzimidazole Schiff base copper (II) complexes as catalysts for environmental and energy applications: VOC oxidation, oxygen reduction and water splitting reactions. International Journal of Hydrogen Energy. 2022;**47**(55):23175-23190. DOI: 10.1016/j.ijhydene.2022.04.271

[16] Verma C, Quraishi MA, Alfantazi A, Rhee KY. Corrosion inhibition potential of chitosan based Schiff bases: Design, performance and applications. International Journal of Biological Macromolecules. 2021;**184**:135-143. DOI: 10.1016/j.ijbiomac.2021.06.049

[17] Yadav M, Sharma S, Devi J. Designing, spectroscopic characterization, biological screening and antioxidant activity of mononuclear transition metal complexes of bidentate Schiff base hydrazones. Journal of Chemical Sciences. 2021;**133**(1):1-22. DOI: 10.1007/s12039-020-01854-6

[18] Singh A, Gogoi HP, Barman P, Guha AK. Novel thioether Schiff base transition metal complexes: Design, synthesis, characterization, molecular docking, computational, biological and catalytic studies. Applied Organometallic Chemistry. 2022;**36**:66-73. DOI: 10.1002/aoc.6673

[19] Dhingra N, Singh JB, Singh HL. Synthesis, spectroscopy, and density functional theory of organotin and organosilicon complexes of bioactive ligands containing nitrogen, sulfur donor atoms as antimicrobial agents: In vitro and in silico studies. Dalton Transactions. 2022;**51**:8821-8831. DOI: 10.1039/D2DT01051H

[20] Abd El-Hamid SM, Sadeek SA, El-Farargy AF, Abd El-Lattif NS. Synthesis, structural characterization and nematicidal studies of some new N2O2 Schiff base metal complexes. Bulletin of the Chemical Society of Ethiopia. 2021;**35**(2):315-335. DOI: 10.4314/bcse.v35i2.12

[21] Alanazi MA, Arafa WA, Althobaiti IO, Altaleb HA, Bakr RB, Elkanzi NA. Green design, synthesis, and molecular docking study of novel Quinoxaline derivatives with insecticidal potential against Aphis craccivora. ACS Omega. 2022;7(31):27674-27689. DOI: 10.1021/acsomega.2c03332

[22] El-Gammal OA, Mohamed FS, Rezk GN, El-Bindary AA. Synthesis, characterization, catalytic, DNA binding and antibacterial activities of Co (II), Ni (II) and Cu (II) complexes with new Schiff base ligand. Journal of Molecular Liquids. 2021;**326**:115223. DOI: 10.1016/j.molliq.2020.115223

[23] Wei L, Zhang J, Tan W, Wang G, Li Q, Dong F, et al. Antifungal activity of double Schiff bases of chitosan derivatives bearing active halogeno-benzenes. International Journal

of Biological Macromolecules. 2021;**179**:292-298. DOI: 10.1016/j.ijbiomac.2021.02.184

[24] Iraji M, Salehi M, Malekshah RE, Khaleghian A, Shamsi F. Liposomal formulation of new arsenic schiff base complex as drug delivery agent in the treatment of acute promyelocytic leukemia and quantum chemical and docking calculations. Journal of Drug Delivery Science and Technology. 2022;**75**:103600. DOI: 10.1016/j.jddst.2022.103600

[25] Hamid SJ, Salih T. Design, synthesis, and anti-inflammatory activity of some Coumarin Schiff Base derivatives: In silico and in vitro study. Drug Design, Development and Therapy. 2022;**16**:2275. DOI: 10.2147/DDDT.S364746

[26] Al-Masoudi NA, Aziz NM, Mohammed AT. Synthesis and in-vitro anti-hiv activity of some new schiff base ligands derived from 5-amino-4- phenyl-4h-1,2,4-triazole-3-thiol and their metal complexes. Phosphorus, Sulfur, and Silicon. 2009;**184**(11):2891-2901. DOI: 10.1080/10426500802591630

[27] Meeran IS, Raja TW, Dusthakeer VA, Ali MM, Tajudeen SS, Shabeer TK. An insight into antimycobacterial and antioxidant potentials of INH-Schiff base complexes and in silico targeting of MtKasB receptor of M. tuberculosis. New Journal of Chemistry. 2022;**46**(10):4620-4633. DOI: 10.1039/D1NJ04977A

[28] Ali MA, Musthafa SA, Munuswamy-Ramanujam G, Jaisankar V. 3-Formylindole-based chitosan Schiff base polymer: Antioxidant and in vitro cytotoxicity studies on THP-1 cells. Carbohydrate Polymers. 2022;**290**:119501. DOI: 10.1016/j.carbpol.2022.119501

[29] Daravath S, Rambabu A, Ganji N, Ramesh G, Lakshmi PA. Spectroscopic, quantum chemical calculations, antioxidant, anticancer, antimicrobial, DNA binding and photo physical properties of bioactive Cu (II) complexes obtained from trifluoromethoxy aniline Schiff bases. Journal of Molecular Structure. 2022;**1249**:131601. DOI: 10.1016/j.molstruc.2021.131601

[30] Mane VA, Palande SV, Swamy DK. In vitro antimicrobial activity and plant growth activity study of schiff base ligand (e)-2, 4-diromo-6-{[(2-(2-methoxyphenoxy) ethyl] iminomethyl} phenol and their complexes with transition metals. Journal of Advanced Scientific Research. 2021;**12**(01 Suppl 2):271-282. DOI: 10.55218/ASR.s12021121sup205

[31] Kaur P, Singh R, Kaur V, Talwar D. Anthranilic acid Schiff base as a fluorescent probe for the detection of arsenite and selenite: A detailed investigation of analytical parameters and mechanism for interaction. Analytical Sciences. 2021;**37**(4):553-560. DOI: 10.2116/analsci.20P102

[32] Refat MS, Saad HA, Gobouri AA, Alsawat M, Adam AM, El-Megharbel SM. Charge transfer complexation between some transition metal ions with azo Schiff base donor as a smart precursor for synthesis of nano oxides: An adsorption efficiency for treatment of Congo red dye in wastewater. Journal of Molecular Liquids. 2022;**345**:117140. DOI: 10.1016/j.molliq.2021.117140

[33] Saha SK, Murmu M, Murmu NC, Banerjee P. Synthesis, characterization and theoretical exploration of pyrene based Schiff base molecules as corrosion inhibitor. Journal of Molecular Structure. 2021;**1245**:131098. DOI: 10.1016/j.molstruc.2021.131098

[34] Zhao ZG, Kodaira T, Nagai N, Hakuta Y, Bando KK, Takashima H,

et al. Self-standing microporous films of arrayed alumina nano-fibers including Schiff base molecules: Effect of the environment around the molecules on their photo-luminescence. Journal of Materials Chemistry. 2012;**22**(19):9738-9744. DOI: 10.1039/C2JM15798E

[35] Yan-Yan LI, Yu-Xi YA, Sha-Sha HO, Yao LI, Zhi YA, Bin-Yu ZH, et al. An electrochemical sensor based on redox-active Schiff Base polymers for simultaneous sensing of glucose and pH. Chinese Journal of Analytical Chemistry. 2021;**49**(6):e21118-e21125. DOI: 10.1016/S1872-2040(21)60107-X

[36] Upendranath K, Venkatesh T, Nayaka YA, Shashank M, Nagaraju G. Optoelectronic, DFT and current-voltage performance of new Schiff base 6-nitro-benzimidazole derivatives. Inorganic Chemistry Communications. 2022;**139**:109354. DOI: 10.1016/j.inoche.2022.109354

[37] Kumar J, Sarma MJ, Phukan P, Das DK. A new simple Schiff base fluorescence "on" sensor for Al 3+ and its living cell imaging. Dalton Transactions. 2015;**44**(10):4576-4581. DOI: 10.1039/C4DT03932G

[38] Spassky N, Wisniewski M, Pluta C, Le Borgne A. Highly stereoelective polymerization of rac-(D, L)-lactide with a chiral schiff's base/aluminium alkoxide initiator. Macromolecular Chemistry and Physics. 1996;**197**(9):2627-2637. DOI: 10.1002/macp.1996.021970902

[39] Kim WS, Choi YK. Electrocatalytic effects of thionyl chloride reduction by polymeric Schiff base transition metal (II) complexes. Applied Catalysis A: General. 2003;**252**(1):163-172. DOI: 10.1016/S0926-860X(03)00414-9

[40] Satheesh CE, Sathish Kumar PN, Kumara PR, Karvembu R, Hosamani A, Nethaji M. Half-sandwich Ru (II) complexes containing (N, O) Schiff base ligands: Catalysts for base-free transfer hydrogenation of ketones. Applied Organometallic Chemistry. 2019;**33**(10):e5111. DOI: 10.1002/aoc.5111

[41] Haque J, Srivastava V, Chauhan DS, Lgaz H, Quraishi MA. Microwave-induced synthesis of chitosan Schiff bases and their application as novel and green corrosion inhibitors: Experimental and theoretical approach. ACS Omega. 2018;**3**(5):5654-5668. DOI: 10.1021/acsomega.8b00455

[42] Nikpassand M, Fekri LZ, Sharafi S. An efficient and green synthesis of novel azo Schiff base and its complex under ultrasound irradiation. Oriental Journal of Chemistry. 2013;**29**(3):1041-1046. DOI: 10.13005/ojc/290326

[43] Shukla M, Kulshrashtha H, Seth DS. Comparative study of the schiff bases by conventional and green method and antimicrobial activity. International Journal of Materials Sciences. 2017;**12**(1):71-76

[44] Sachdeva H, Saroj R, Khaturia S, Dwivedi D. Operationally simple green synthesis of some Schiff bases using grinding chemistry technique and evaluation of antimicrobial activities. Green Processing and Synthesis. 2012;**1**(5):469-477. DOI: 10.1515/gps-2012-0043

[45] Kargar H, Fallah-Mehrjardi M, Behjatmanesh-Ardakani R, Torabi V, Munawar KS, Ashfaq M, et al. Sonication-assisted synthesis of new Schiff bases derived from 3-ethoxysalicylaldehyde: Crystal structure determination, Hirshfeld surface analysis, theoretical calculations and spectroscopic studies. Journal of Molecular Structure. 2021;**1243**:130782. DOI: 10.1016/j.molstruc.2021.130782

[46] Mittersteiner M, Farias FF, Bonacorso HG, Martins MA, Zanatta N. Ultrasound-assisted synthesis of pyrimidines and their fused derivatives: A review. Ultrasonics Sonochemistry. 2021;**79**:105683. DOI: 10.1016/j.ultsonch.2021.105683

[47] Otani N, Furuya T, Katsuumi N, Haraguchi T, Akitsu T. Synthesis of amino acid derivative Schiff base copper (II) complexes by microwave and wet mechanochemical methods. Journal of the Indian Chemical Society. 2021;**98**(1):100004. DOI: 10.1016/j.jics.2021.100004

[48] Meena DR, Aalam MJ, Chaudhary P, Yadav GD, Singh S. Synthesis and structural studies of Pd (II) complexes of bidentate Schiff bases and their catalytic activities as pre-catalysts in the Mizoroki-heck reaction. Polyhedron. 2022;**222**:115931. DOI: 10.1016/j.poly.2022.115931

[49] Alfonso-Herrera LA, Rosete-Luna S, Hernández-Romero D, Rivera-VillanuevaJM, Olivares-RomeroJL, Cruz-Navarro JA, et al. Transition metal complexes with tridentate Schiff bases (O^ N^ O and O^ N^ N) derived from salicylaldehyde: An analysis of their potential anticancer activity. ChemMedChem. 2022;**17**:1-47. DOI: 10.1002/cmdc.202200367

[50] Abdel-Rahman LH, Adam MS, Al-Zaqri N, Shehata MR, Ahmed HE, Mohamed SK. Synthesis, characterization, biological and docking studies of ZrO (II), VO (II) and Zn (II) complexes of a halogenated tetra-dentate Schiff base. Arabian Journal of Chemistry. 2022;**15**(5):103737. DOI: 10.1016/j.arabjc.2022.103737

[51] Das S, Das M, Laha S, Rajak K, Choudhuri I, Bhattacharyya N, et al. Development of moderately fluorescence active salen type chemosensor for judicious recognition and quantification of Zn (II), Al (III) and SO4=: Demonstration of molecular logic gate formation and live cell images studies. Journal of Molecular Structure. 2022;**1263**:133214. DOI: 10.1016/j.molstruc.2022.133214

[52] Li Y, Liu Y, Wei X, Wang L, Wang Y, Zhang Q. A symmetric Schiff base fluorescent "turn-on" chemosensor for aluminum (III) ion selective detection based on hydrolysis. Journal of Molecular Structure. 2022;**1261**:132884. DOI: 10.1016/j.molstruc.2022.132884

[53] Chaudhary NK, Guragain B. Coordination chemistry, antibacterial screening, and In silico ADME study of mononuclear NiII and CuII complexes of asymmetric Schiff Base of streptomycin and aniline. Journal of Chemistry. 2022;**2022**:1-14. DOI: 10.1155/2022/3881217

[54] Yusof EN, Latif MA, Tahir MI, Sakoff JA, Veerakumarasivam A, Page AJ, et al. Homoleptic tin (IV) compounds containing tridentate ONS dithiocarbazate Schiff bases: Synthesis, X-ray crystallography, DFT and cytotoxicity studies. Journal of Molecular Structure. 2020;**1205**:127635. DOI: 10.1016/j.molstruc.2019.127635

[55] Dar OA, Lone SA, Malik MA, Wani MY, Talukdar MI, Al-Bogami AS, et al. Heteroleptic transition metal complexes of Schiff-base-derived ligands exert their antifungal activity by disrupting membrane integrity. Applied Organometallic Chemistry. 2019;**33**(11):e5128. DOI: 10.1002/aoc.5128

[56] Abdel-Rahman LH, Abu-Dief AM, Adam MS, Hamdan SK. Some new nano-sized mononuclear Cu (II) Schiff base complexes: Design, characterization, molecular modeling and catalytic

potentials in benzyl alcohol oxidation. Catalysis Letters. 2016;**146**(8):1373-1396. DOI: 10.1007/s10562-016-1755-0

[57] Parasuraman B, Rajendran J, Rangappan R. An insight into antibacterial and anticancer activity of homo and hetero binuclear Schiff Base complexes. Oriental Journal of Chemistry. 2017;**33**(3):1223. DOI: 10.13005/OJC/330321

[58] Puzari A, Borah D, Das P. Binuclear Pd (II) complexes with multidentate Schiff base ligands: Synthesis, catalysis, and antibacterial properties. Monatshefte für Chemie-Chemical Monthly. 2022;**153**:435-442. DOI: 10.1007/s00706-022-02929-5

[59] Maiti SK, Bora A, Barman P, Chandra AK, Baidya B. Binuclear chiral Ni (II) complex of tridentate OON chiral Schiff base ligand, 1-((E)-(((1S, 2R)-2-hydroxy-2, 3-dihydro-1H-inden-1-yl) imino) methyl) naphthalen-2-ol. Journal of Coordination Chemistry. 2020;**73**(1):67-86. DOI: 10.1080/00958972.2019.1711069

[60] Dey D, Kaur G, Ranjani A, Gayathri L, Chakraborty P, Adhikary J, et al. A trinuclear zinc–schiff base complex: Biocatalytic activity and cytotoxicity. European Journal of Inorganic Chemistry. 2014;**2014**(21):3350-3358. DOI: 10.1002/ejic.201402158

[61] Mokolokolo PP, Frei A, Tsosane MS, Kama DV, Schutte-Smith M, Brink A, et al. Nuclearity manipulation in Schiff-base fac-tricarbonyl complexes of Mn (I) and Re (I). Inorganica Chimica Acta. 2018;**471**:249-256. DOI: 10.1016/j.ica.2017.10.036

[62] Tang J, Yao P, Huang F, Luo M, Wei Y, Bian H. Stereoselective sulfoxidation catalyzed by achiral Schiff base complexes in the presence of serum albumin in aqueous media. Tetrahedron: Asymmetry. 2017;**28**(12):1700-1707. DOI: 10.1016/j.tetasy.2017.10.021

[63] Roy GB. Synthesis and study of physico-chemical properties of a new chiral Schiff base ligand and its metal complex. Inorganica Chimica Acta. 2009;**362**(6):1709-1714. DOI: 10.1016/j.ica.2008.08.009

[64] Ito M, Akitsu T, Palafox MA. Theoretical interpretation of polarized light-induced supramolecular orientation on the basis of normal mode analysis of azobenzene as hybrid materials in PMMA with chiral Schiff base Ni (II), Cu (II), and Zn (II) complexes. Journal of Applied Solution Chemistry and Modeling. 2016;**5**(1):30-47. DOI: 10.6000/1929-5030.2016.05.01.3

[65] Sunaga N, Haraguchi T, Akitsu T. Orientation of chiral schiff base metal complexes involving azo-groups for induced cd on gold nanoparticles by polarized uv light irradiation. Symmetry. 2019;**11**(9):1094. DOI: 10.3390/sym11091094

[66] Alshaheri AA, Tahir MI, Rahman MB, Begum T, Saleh TA. Synthesis, characterisation and catalytic activity of dithiocarbazate Schiff base complexes in oxidation of cyclohexane. Journal of Molecular Liquids. 2017;**240**:486-496. DOI: 10.1016/j.molliq.2017.05.081

[67] Gupta KC, Sutar AK. Catalytic activities of Schiff base transition metal complexes. Coordination Chemistry Reviews. 2008;**252**(12-14):1420-1450. DOI: 10.1016/j.ccr.2007.09.005

[68] Abuamer KM, Maihub AA, El-Ajaily MM, Etorki AM, Abou-Krisha MM, Almagani MA. The role

of aromatic Schiff bases in the dyes techniques. International Journal of Organic Chemistry. 2014;**4**:7-15. DOI: 10.4236/ijoc.2014.41002

[69] Kazemnejadi M, Dehno Khalaji A, Mighani H. Synthesis and characterization of Schiff-base polymer derived from 2, 5-dichloroaniline and 2-hydroxybenzaldehyde. Quarterly Journal of Iranian Chemical Communication. 2017;**5**(3):237-363. Serial No. 16):237-41

[70] Higueras L, López-Carballo G, Gavara R, Hernández-Muñoz P. Reversible covalent immobilization of cinnamaldehyde on chitosan films via schiff base formation and their application in active food packaging. Food and Bioprocess Technology. 2015;**8**(3):526-538. DOI: 10.1007/s11947-014-1421-8

[71] Mr SN, Srinivasan AK, Keerthana P, Kumar A. Schiff's base (SB) modified zirconium dioxide reinforced PLA bio-composite film for industrial packaging applications. Composites Communications. 2021;**25**:100750. DOI: 10.1016/j.coco.2021.100750

[72] Natesan S, Samuel JS, Srinivasan AK. Design and development of Schiff's base (SB)-modified polylactic acid (PLA) antimicrobial film for packaging applications. Polymer Bulletin. 2022;**79**(7):4627-4646. DOI: 10.1007/s00289-021-03703-z

[73] Singh S. Synthesis, spectroscopic studies and pesticidal activity of transition metal complexes with unsymmetrical schiff base. Indian Journal of Biochemistry and Biophysics (IJBB). 2021;**58**(6):565-571. DOI: 10.56042/ijbb.v58i6.57792

[74] Karmakar I, Mandal S, Mitra A. Evaluation of antimicrobial and insect repellent properties of two novel zinc (II), and nickel (II) complexes containing a tetradentate Schiff Base. Journal of Integrated Science and Technology. 2015;**3**(2):60-67

[75] Kapoor P, Singh RV, Fahmi N. Coordination chemistry of rare earth metal complexes with coumarin-based imines: Ecofriendly synthesis, characterization, antimicrobial, DNA cleavage, pesticidal, and nematicidal activity evaluations. Journal of Coordination Chemistry. 2012;**65**(2): 262-277. DOI: 10.1080/00958972.2011.649265

[76] Fakhari AR, Khorrami AR, Naeimi H. Synthesis and analytical application of a novel tetradentate N2O2 Schiff base as a chromogenic reagent for determination of nickel in some natural food samples. Talanta. 2005;**66**(4):813-817. DOI: 10.1016/j.talanta.2004.12.043

[77] Aksuner N, Henden E, Yilmaz I, Cukurovali A. A highly sensitive and selective fluorescent sensor for the determination of copper (II) based on a schiff base. Dyes and Pigments. 2009;**83**(2):211-217. DOI: 10.1016/j.dyepig.2009.04.012

[78] López-Herraiz M, Castillo-Martínez E, Carretero-González J, Carrasco J, Rojo T, Armand M. Oligomeric-Schiff bases as negative electrodes for sodium ion batteries: Unveiling the nature of their active redox centers. Energy & Environmental Science. 2015;**8**(11):3233-3241. DOI: 10.1039/C5EE01832C

[79] Yang X, Zhuang X, Huang Y, Jiang J, Tian H, Wu D, et al. Nitrogen-enriched hierarchically porous carbon materials fabricated by graphene aerogel templated Schiff-base chemistry for high performance electrochemical capacitors. Polymer Chemistry. 2015;**6**(7):1088-1095. DOI: 10.1039/C4PY01408A

[80] Castillo-Martínez E, Carretero-González J, Armand M. Polymeric Schiff bases as low-voltage redox centers for sodium-ion batteries. Angewandte Chemie International Edition. 2014;**53**(21):5341-5345. DOI: 10.1002/anie.201402402

[81] Chen H, Zhang Y, Xu C, Cao M, Dou H, Zhang X. Two π-conjugated covalent organic frameworks with long-term Cyclability at high current density for lithium ion battery. Chemistry. A European Journal. 2019;**25**(68):15472-15476. DOI: 10.1002/chem.201903733

[82] Oiye ÉN, Ribeiro MF, Katayama JM, Tadini MC, Balbino MA, Eleotério IC, et al. Electrochemical sensors containing schiff bases and their transition metal complexes to detect analytes of forensic, pharmaceutical and environmental interest. A review. Critical Reviews in Analytical Chemistry. 2019;**49**(6):488-509. DOI: 10.1080/10408347.2018.1561242

[83] Wei W, Liu Z, Liang C, Han GC, Han J, Zhang S. Synthesis, characterization and corrosion inhibition behavior of 2-aminofluorene bis-Schiff bases in circulating cooling water. RSC Advances. 2020;**10**(30):17816-17828. DOI: 10.1039/D0RA01903H

[84] Krishnan U, Iyer SK. Iminothiophenol Schiff base-based fluorescent probe for dual detection of Hg2+ and Cr3+ ions and its application in real sample analysis. Journal of Photochemistry and Photobiology A: Chemistry. 2022;**425**:113663. DOI: 10.1016/j.jphotochem.2021.113663

[85] Bharali B, Talukdar H, Phukan P, Das DK. A new Schiff Base based fluorescent sensor for Al (III) based on 2-Hydroxyacetophenone and o-Phenylenediamine. Journal of Fluorescence. 2020;**30**(4):751-757. DOI: 10.1007/s10895-020-02527-w J

[86] Yuan C, Liu X, Wu Y, Lu L, Zhu M. A triazole Schiff base-based selective and sensitive fluorescent probe for Zn2+: A combined experimental and theoretical study. Spectrochimica Acta Part A: Molecular and Biomolecular Spectroscopy. 2016;**154**:215-219. DOI: 10.1016/j.saa.2015.10.035

[87] Fatima B, Hussain D, Bashir S, Hussain HT, Aslam R, Nawaz R, et al. Catalase immobilized antimonene quantum dots used as an electrochemical biosensor for quantitative determination of H2O2 from CA-125 diagnosed ovarian cancer samples. Materials Science and Engineering: C. 2020;**117**:111296. DOI: 10.1016/j.msec.2020.111296

[88] Abou-Omar MN, Attia MS, Afify HG, Amin MA, Boukherroub R, Mohamed EH. Novel optical biosensor based on a nano-gold coated by Schiff base doped in sol/gel matrix for sensitive screening of oncomarker CA-125. ACS Omega. 2021;**6**(32):20812-20821. DOI: 10.1021/acsomega.1c01974

[89] Sheta SM, Akl MA, Saad HE, El-Gharkawy ES. A novel cerium (iii)–isatin Schiff base complex: Spectrofluorometric and DFT studies and application as a kidney biomarker for ultrasensitive detection of human creatinine. RSC Advances. 2020;**10**(10):5853-5863. DOI: 10.1039/C9RA10133K

[90] Fahmi N, Shrivastava S, Meena R, Joshi SC, Singh RV. Microwave assisted synthesis, spectroscopic characterization and biological aspects of some new chromium (iii) complexes derived from N⁁O donor Schiff bases. New Journal of Chemistry. 2013;**37**(5):1445-1453. DOI: 10.1039/C3NJ40907D

[91] Sharma S, Meena R, Singh RV, Fahmi N. Synthesis, characterization, antimicrobial, and DNA cleavage evaluation of some organotin (IV) complexes derived from ligands containing the 1H-indole-2, 3-dione moiety. Main Group Metal Chemistry. 2016;**39**(1-2):31-40. DOI: 10.1515/mgmc-2015-0030

[92] Kumari A, Meena R, Singh RV, Fahmi N. Synthesis, characterization, antimicrobial and DNA cleavage study of organoantimony (III) and organoarsenic (III) complexes with monofunctional bidentate Schiff base. Indian Journal of Chemistry. 2021;**60**:341-347

Chapter 5

Some Aromatic Schiff Bases and Their Metal Complexes

Mirjana M. Radanović and Berta Barta Holló

Abstract

Schiff bases represent a widely studied group of compounds, interesting from the synthetic and structural, but also applicational point of view. Among these, aromatic Schiff bases stand out due to their stability and possible application as safer therapeutic agents. Since the presence of the phenolic group and the heteroatom in the benzene ring seems to be of interest for the biological activity of Schiff bases, imine derivatives of some dihydroxybenzaldehydes, 2-acetylpyridine, and 2,6-diacetylpyridine are chosen to be described here. The syntheses, structural features, as well as investigations of possible application of Schiff bases themselves, but also their metal complexes are described. The Cambridge Structural Database has been searched to obtain a better knowledge of the structures of these Schiff bases and their metal complexes, structural changes upon coordination, and the most common and interesting modes of coordination. The main accent is given to those compounds that are promising candidates for medicinal and pharmaceutical use, but some of the described complexes are good as catalysts, optical materials, sensors, analytical reagents, etc.

Keywords: dihydroxybenzaldehydes, 2-acetylpyridine, 2,6-diacetylpyridine, imine derivative, metal complexes, synthesis, crystal structure, CSD search

1. Introduction

Nowadays, one of the greatest imperatives is the synthesis and characterization of new drugs with enhanced activity and safer application [1]. One of the promising groups of compounds in this field is undoubtedly Schiff bases and their metal complexes, due to the ease of synthesis, interesting coordination properties, and wide range of possible applications [2–4]. Research has shown that aromatic carbonyl compounds form stable Schiff bases, while those with aliphatic compounds are prone to polymerization. Also, aromatic derivatives of some therapeutic agents were shown to be safer to use [5].

Having in mind the bond between biological activity and complexation ability [2], thorough examinations have been done to explore the possibilities of syntheses of metal complexes with aromatic Schiff bases, as well as the properties of the obtained compounds. The overview, synthesis, and comparison of some structural features of all so far known metal complexes with selected Schiff bases of some dihydroxybenzaldehydes, 2-acetylpyridine, and 2,6-diacetylpyridine are presented.

2. Schiff bases of dihydroxybenzaldehydes

In this part, the significance, structure, and potential application of the Schiff bases of 2,3-; 2,4-; and 2,5-dihydroxybenzaldehyde (**Figure 1**) and their metal complexes will be presented.

Since the knowledge about the structure of compounds leads to better insight into the mechanism of their activities, thus the prediction of improvements, the search of Cambridge Structural Database (CSD) [6] about the number and structural features of these Schiff bases and their metal complexes has been made. The comparative analysis of structural properties of free Schiff bases and coordinated ones is very important for their future applications.

Imine derivatives of 2,3-; 2,4-; and 2,5-dihydroxybenyzaldehyde are part of the larger group of *o*-hydroxy Schiff bases. These compounds are interesting from the structural point of view, due to their ability to form intramolecular H-bonds. Also, metal complexes with Schiff bases containing phenol group through which the coordination is realized are considered models for important biological reactions, which mimic catalytic behavior of metalloenzymes [7]. Macrocyclic derivatives of these Schiff bases are the basis for the investigation of many biological processes, such as photosynthesis and oxygen metabolism [8].

However, their existence in two different tautomeric forms, that is, phenol-imines and keto-amines (**Figure 2**), is one of the crucial factors in the determination of their photochromic and thermochromic properties [9]. Based on the tautomeric form, two types of intramolecular H-bonds could be formed: O−H···N in phenol-imines and N−H···O in keto-amines [9–11]. Another form of Schiff bases is zwitterionic tautomer, which is rather rare compared with the previously described [12]. This is in accordance with our findings based on the search of CSD.

Another interesting feature of dihydroxybenzaldehydes is the presence of another OH-group, which enables the formation of more H-bonds, but also the additional coordination *via* another oxygen atom (*vide infra*).

The CSD search has shown that 75 mono-condensed, 18 bi-condensed, two tri-condensed, and two tetra-condensed Shiff bases of 2,3-dihydroxybenzaldehyde have been structurally characterized. The number of structures of the Schiff bases with 2,4-dihydroxybenzaldehyde is similar (79 mono-condensed and 8 bi-condensed), while the number of imine derivatives of 2,5-dihydroxybenzaldehyde is significantly lower—27 structures of mono-condensed and only one structure of bi-condensed Schiff bases. In all of those structures, dihydroxybenzaldehyde fragment is present in its neutral form. As it is said before, the zwitterionic form is rare—29 structures of Schiff bases of 2,3-DHBA, but only 9 and 4 structures of Schiff bases of 2,4- and 2,5-DHBA, respectively. However, there is a certain number of structures in which

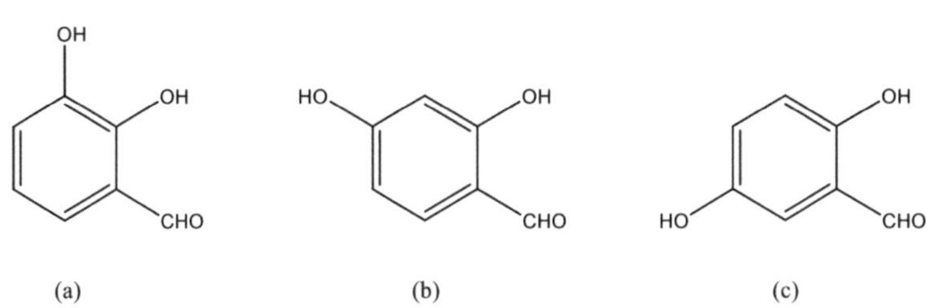

Figure 1.
Structural formulas of 2,3- (a), 2,4- (b), and 2,5-DHBA (c).

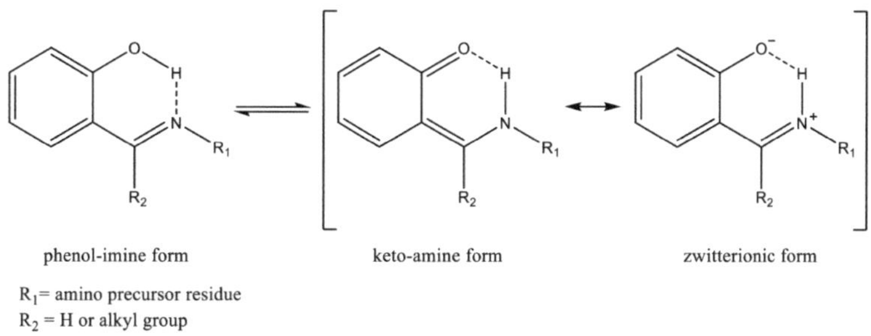

Figure 2.
Tautomerism in o-hydroxy Schiff bases.

two dihydroxybenzaldehyde moieties are present, that isthat is, one in neutral form and the other one in its monodeprotonated form. In the latter, the migration of H-atom from hydroxy-group in position 2 to azomethine nitrogen atom takes place and results in the formation of zwitterion of the Schiff base. This leads to changes in bond lengths, that is, the shortening of the C—O bond for ca. 0.06–0.07 Å, and the elongation of the C—N bond for ca. 0.02 Å.

One of the main areas of possible applications of Schiff bases, in general, is medicine or pharmacology. From this point of view, the most important characteristic of Schiff bases and an essential structural requirement for biological activity is their ability to cleave DNA [13]. The condensation of 2,3-, and 2,4-DHBA with tert-butylamine described in [14] gave two types of compounds. The analysis of the binding of these two ligands to human serum albumin (HSA) indicates the dominance of the latter. H-bonds stabilize the albumin-ligand system and have the main role in this type of binding.

Almost half a century ago, attempts for finding a cure for sickle cell anemia included the investigation of Schiff bases formed by amino groups of intracellular hemoglobin with aromatic aldehydes, such as 2,3-DHBA. Hydroxyl group as a substituent can lead 2,3-DHBA to specific hemoglobin sites, important for oxy-deoxy equilibrium. In this way, the affinity of sickle cells toward oxygen is increased, thus can lead to the mitigation of the effects of this disease [15].

The search of CSD revealed 136 structures of complex compounds with Schiff bases of 2,3-DHBA with 3d-, 4d-, 4f-, and s-metals, as well as uranium and organotin compounds. In the largest number of these complexes, Schiff bases have the role of tridentate (33 structures) (**Figure 3a**) and tetradentate ligand (**Figure 4a** and **b**) (20 structures). Significantly less common are bidentate (9 structures), pentadentate (2 structures), and hexadentate (2 structures) coordination of a chelating ligand. Due to the existence of two neighboring hydroxy groups, the coordination modes of these ligands could be described as versatile. In ca. 50 structures, additional bridging coordination of oxygen atom of one or both hydroxy groups is found (**Figure 3e**). Structures of numerous mixed complexes with these ligands containing 4f-metal centers are determined. A series of four mixed complexes with palladium and 4f-metals stands out. In these complexes besides the tetradentate ONNO coordination of the Schiff base, four oxygen donors of the neighboring OH-groups are coordinated to another metal center (**Figure 4d**). Also, one organometallic compound of palladium(II) is structurally characterized [16].

Complexes of 2,4-DHBA Schiff bases containing 3d-, 4d-metals, organotin compounds, lead and uranium are structurally characterized. The most common

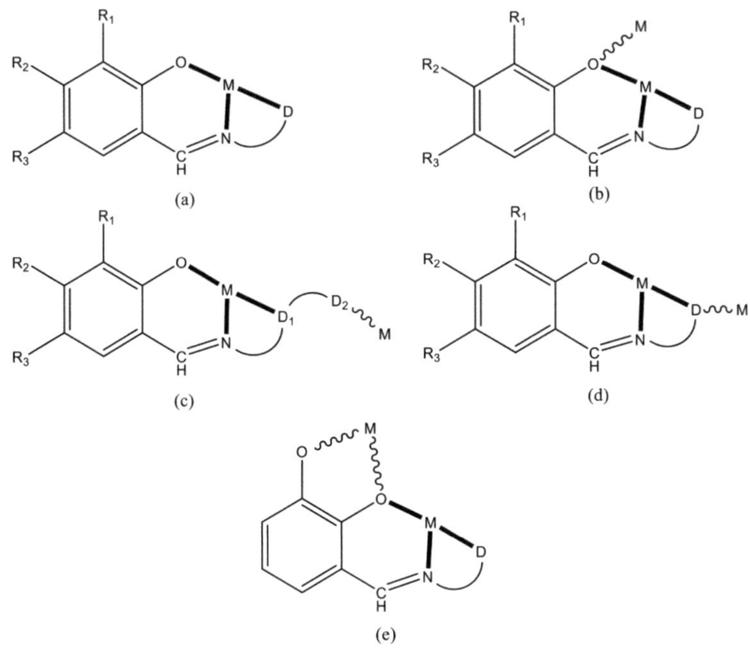

Figure 3.
The most common modes of tridentate (a) and additional bridging coordination (b-e) of Schiff bases of 2,3-DHBA (R_1 = OH, R_2 = R_3 = H), 2,4-DHBA (R_2 = OH, R_1 = R_3 = H), and 2,5-DHBA (R_3 = OH, R_1 = R_2 = H).

coordination modes are ONX (X = O, N, S) tridentate (43 structures) (**Figure 3a**) and ONNO, ONNN, and ONOO tetradentate (42 structures). Unlike those, there are only 11 and 14 structures with bidentate and hexadentate coordination of these ligands, respectively. In a certain number of structures, bridging coordination of these Schiff bases is found, that is, in 8 structures besides the three donor atoms, an additional coordination bond is realized through some atom of the imine residue (**Figure 3c** and **d**), while in the other 8 structures, the bridging atom is deprotonated phenolic oxygen from the position 2 (**Figure 3b**). In the other 8 structures, both of those bridging coordination modes are proven. Finally, there are 5 structures with tetradentate and additional bridging coordination of another donor atom, as well as two structures with bidentate coordinated Schiff base and bridging of an oxygen atom from position 2 of the 2,4-DHBA moiety.

On the contrary, there are far fewer structures of the complexes with Schiff basses of 2,5-DHBA—only 12 complexes with 3d-metals and organotin. Nevertheless, the coordination modes of these ligands are numerous—in two structures bidentate coordination mode is found, in three structures—tridentate, in four structures—tetradentate, and in three structures additional bridging coordination of phenolic oxygen atom and/or another donor atom of imine residue (**Figure 3c–e**).

Unlike OH-group from position 3 in 2,3-DHBA residue, in Schiff bases of 2,4-DHBA and 2,5-DHBA additional hydroxy groups (from positions 4 and 5, respectively) do not take part in coordination.

In the silver(I) complex (Ref. code MICVUS) [17] the organic ligand is coordinated in an exotridentate manner to three metal ions (**Figure 5**). It is interesting to note that the protonated hydroxy group of 2,4-DHBA residue is involved in coordination.

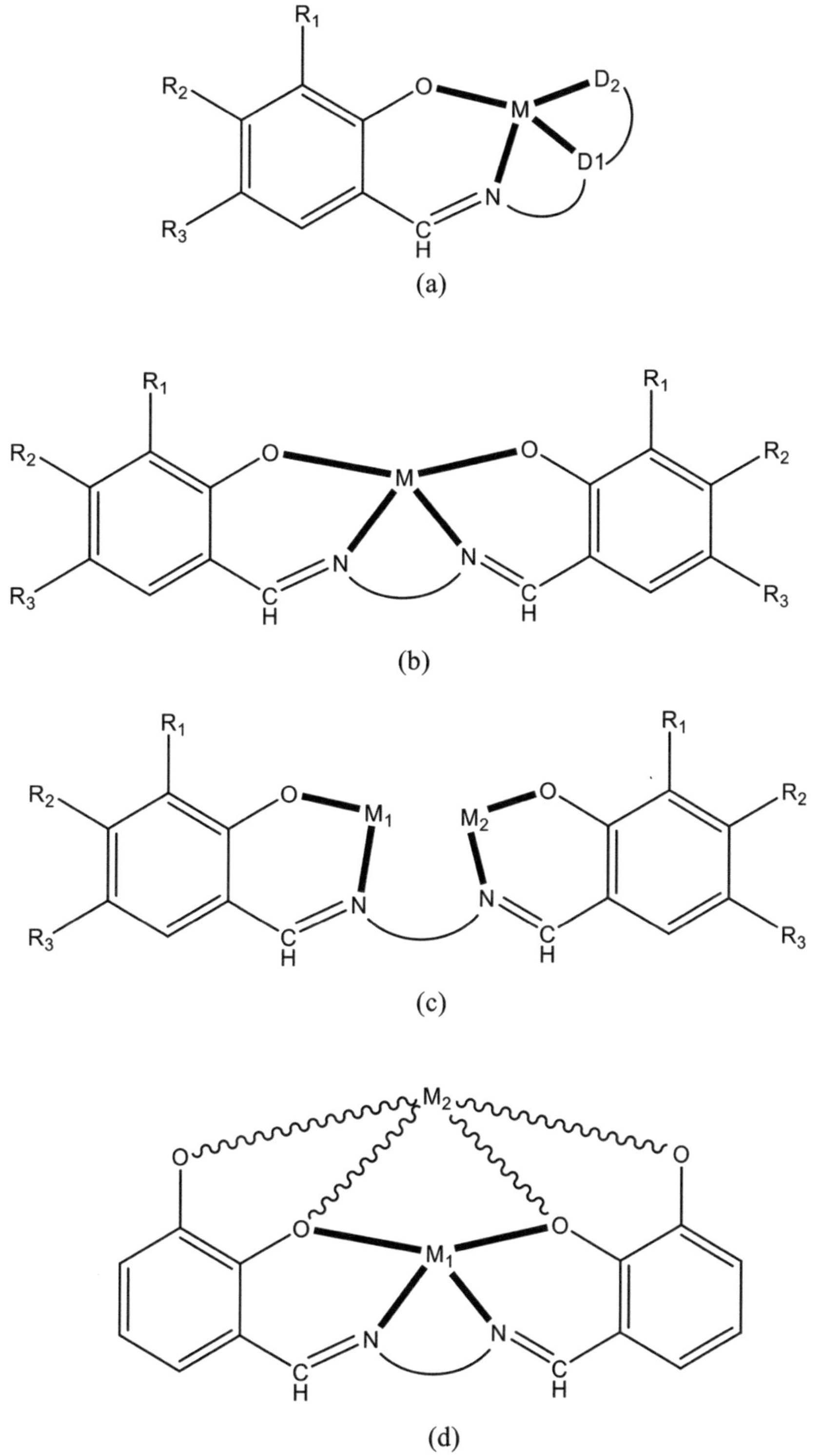

Figure 4.
The most common modes of tetradentate (a-c) and additional bridging coordination (d) of Schiff bases of 2,3-DHBA (R_1 = OH, R_2 = R_3 = H), 2,4-DHBA (R_2 = OH, R_1 = R_3 = H), and 2,5-DHBA (R_3 = OH, R_1 = R_2 = H).

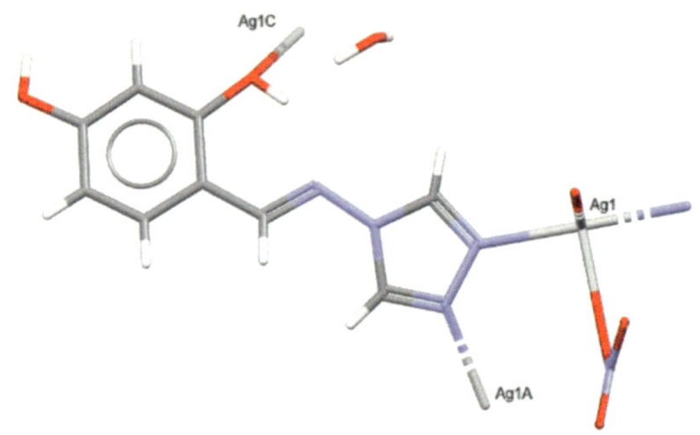

Figure 5.
Molecular structure of Ag(I) complex with the Schiff base of 2,4-DHBA.

Dihydroxybenzaldehyde residue does not take part in coordination in two ferrocene derivatives (Ref. codes OWUXUB [18] and RUTYIP [19]), as well as in three Cu(I) complexes (Ref. codes VAHHUL, VAHJIB, and VAHKIC) [20] and one silicone complex (Ref. code XOSMOJ) in which the coordination is realized through the residue of the imine ligand precursor [21].

Besides, in one cadmium(II) complex (Ref. code QEWFUW) [22], the Schiff base of 2,4-DHBA is present in its neutral form but is not involved in coordination. Also, in two complexes of potassium and calcium (Ref. codes MAVNEE and MAVNII) [23] coordination is accomplished *via* sulfonato oxygen atoms, without the involvement of a 2,4-DHBA fragment. This is a good example of Pearson's theory of hard and soft acids and bases.

Schiff bases of 2,3-DHBA and their metal complexes showed to be important in the area of supramolecular chemistry, as precursors for obtaining octanuclear cluster compounds of zinc(II) [24], and some of them show significant photoluminescence [25]. Bidentate Schiff base of 2,3-DHBA and alkylamines is reported to be a good chelating ligand with significant selectivity and sensitivity for Co(II), Cu(II), and Fe(II) ions [26]. The asymmetric Schiff base obtained by condensation of allylamine has good antimicrobe activity, even better than its complex with Mo(VI) [27].

Octahedral complexes of chromium(III) with Schiff bases of metformin and 2,3-DHBA, 2,4-DHBA, 2,5-DHBA, and 3,4-DHBA are obtained by template reactions. For the complex with 2,4-DHBA Schiff base, $[CrLCl(H_2O)_2]\cdot 3H_2O$, the research revealed the potential effect in lowering blood glucose levels [28].

A structurally interesting compound is the complex of zinc(II) with the Schiff base of 2,3-DHBA and aminoguanidine, of the formula $[Zn_2L(CH_3COO)_2]_2\cdot 2MeCN$ (**Figure 6**). The asymmetric unit consists of the dianion of the chelating ligand, which is in an ONN tridentate manner coordinated to one metal center, and in a monodentate manner (*via* the oxygen atom of the deprotonated hydroxy group from position 3) to another zinc(II) ion, two acetate ions that have a bridging role, and one solvent molecule [29].

The research has shown that the tridentate Schiff base of 2,4-DHBA and aminoguanidine expresses antioxidative and antiglycation effects, similar to the Schiff base of pyridoxal and aminoguanidine. However, the Schiff base with 2,4-DHBA had a better *in vivo* antithrombic effect and thus could be useful in the treatment of diabetic complications [30]. One square-planar complex of copper(II) with this Schiff base has been synthesized and structurally characterized (**Figure 7**). In this complex, the chelating ligand is coordinated through the oxygen atom of the deprotonated hydroxy

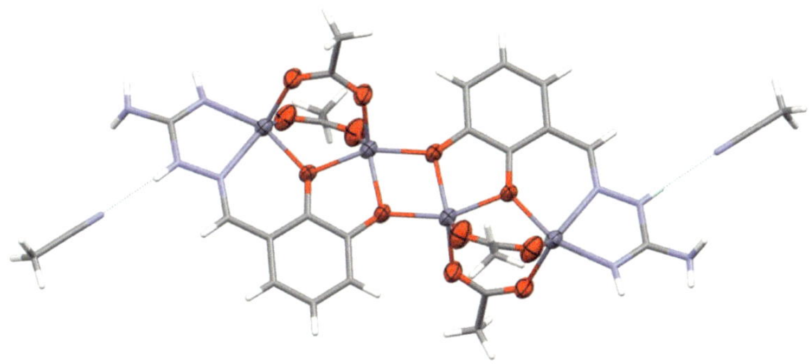

Figure 6.
Molecular structure of $[Zn_2L(CH_3COO)_2]_2{\cdot}2MeCN$.

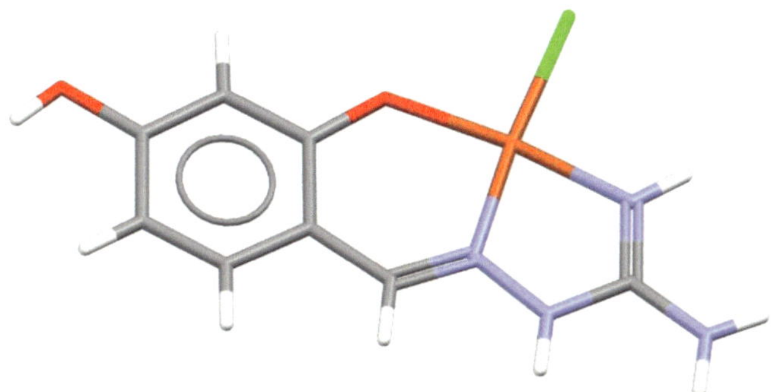

Figure 7.
Molecular structure of the complex $[Cu(L\text{-}H)Cl]{\cdot}nH_2O$ ($n = 4$) (water molecules are not shown).

group from position 2, the nitrogen atom of the azomethine group, and the nitrogen atom of the imine group of the aminoguanidine fragment [31].

3. Schiff bases of 2-acetylpyridine

3.1 Significance and application

Schiff bases containing N donors and their metal complexes showed to be promising candidates for fighting against drug resistance, which many microorganisms have developed toward the drugs used [32, 33]. One of the good precursors for the synthesis of potentially biologically active imine derivatives is 2-acetylpyridine. Besides the potential for medicinal application, Schiff bases of this ketone have a wide range of other uses, such as fluorescent and luminescent probes [34, 35], optical sensors for determination of cadmium(II) in natural water and art-paint samples [36], effective catalysts [37], etc. The research has shown that imine derivatives of 2-acetylpyridine and their metal complexes express substantial in vitro activity against various human tumor lines [38, 39] and significant cytotoxicity against some solid tumors and leukemia cells [40].

Some cobalt(III) complexes with cytotoxic 2-acetylpyridine-derived hydrazones can release the bioactive ligand upon reduction. Also, these complexes interacted with HSA, indicating that they could be transported by this protein [40].

Complexation of N(4)-tolyl-2-acetylpyridine-derived thiosemicarbazones to gold(III), platinum(II), and platinum(IV) enhanced cytotoxic activity of the Schiff base [41]. Research also showed that bismuth(III), antimony(III), gallium(III), palladium(II), platinum(II), and tin(IV) complexes with 2-acetylpyridine-N(4)-ortho-chlorophenylthiosemicarbazone express cytotoxic activity to some solid tumor cells and that gallium(III), bismuth(III) and rhenium(I) complexes with 2-acetyl-pyridine- and 2-benzoylpyridine-derived hydrazones express cytotoxic effects against solid tumor and leukemia cells [42].

Recently, some lanthanide(III) complexes with 2-acetylpyridine derivatives were suggested to have great potential in biomedical application, due to their ability to inhibit proliferation and induce apoptosis in human lung and gastric cancer cell lines, thus good antitumor activities [43].

Due to the possible biologic activity of Schiff bases of amino acids and their metal complexes, it is not surprising that they represent a very interesting field of investigation [44]. Some copper(II) complexes with Schiff base of 2-acetylpyridine and leucine showed significant analgesic and antirheumatic effects, while the Schiff base with L-tryptophan forms 3D coordination polymers with copper(II), zinc(II), and cadmium(II) that express antitumor activity toward breast cancer cells [45]. Interestingly, the most promising of those is the cadmium complex, which showed the highest antiproliferative activity and potential to inhibit proteasomal chymotrypsin-like activity and induce apoptosis of human breast cancer cell lines.

In [2] some 3d-metal complexes are described as potential antibacterial drugs that could be used to control some Gram+ bacteria nosocomial infections in the hospital. Another series of 3d-metal complexes of Schiff base of 2-acetylpyridine and 4,4-oxy-dianniline express not only antibacterial but also anticorrosion behavior [46].

Some 2-acetylpyridine derivatives and their gallium(III) complexes were proposed to be good antitumor drug candidates to treat brain tumors [47].

The most recent research [48] describes synthesis and structure of dinuclear silver(I) complex with selective anticancer activity. This compound inhibits the growth of cancer cells, while it does not make adverse effects on normal cells.

Since the biological activity is in the narrow bond with the structure of the compounds, some biologically active and, at the same time, structurally interesting complexes will be described in the following section.

3.2 Coordination behavior

The survey of CSD revealed 132 structures of free Schiff bases of 2-acetylpyridine. Most of these compounds are neutral, and only the sixth part of the total number has protonated pyridine nitrogen atom, thus exists in the form of salts or zwitterionic form. The number of complexes with 2-acetylpyridine imine derivatives is significantly larger, that is, almost 1400 structures. Analysis of these data leads to the conclusion that the denticity of those compounds varies from monodentate to heptadentate ligands. There are examples of structures in which pyridine or azomethine nitrogen atoms do not take part in coordination. However, monodentate mode always includes coordination of donor atom of amine precursor residue, usually sulfur.

It is interesting to mention that there are complexes in which two moieties of the Schiff base are coordinated differently. For example, in Pt(II) complex with 2-acetyl-pyridine-4 N-ethyl thiosemicarbazone (**Figure 8**, Ref. code ESAZIH) one ligand moiety is coordinated as NNS tridentate, while the other one achieves coordination *via* sulfur atom of thiosemicarbazide residue [49]. In this complex, the tridentate

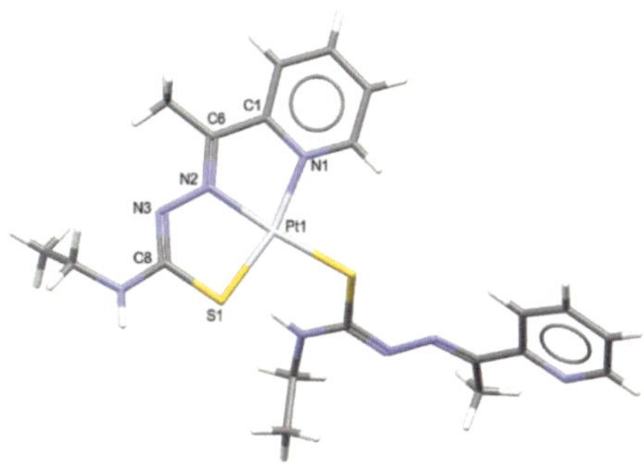

Figure 8.
Molecular structure of Pt(II) complex with 2-acetylpyridine-4N-ethyl thiosemicarbazone.

ligand shows a Z, E, Z configuration about the C1−C6, C6−N2, and N3−C8 bonds for the two nitrogen and one sulfur donor atoms, while the monodentate ligand shows an E, E, Z configuration for the same atoms.

The described mode of coordination could be governed by many factors, such as rigid ligands, preferences of the metal toward the sulfur as softer base over nitrogen, stability of the tricyclic ring system, as well as preferences of Pt(II) to be in square-planar environments [49]. This complex is synthesized as a part of a series of Pt(II) and Pd(II) complexes, which showed to be promising candidates to overcome cis-platin resistance. The same coordination modes, as well as the same configuration of ligands, were found in other two Pd(II) complexes (Ref. codes MUSVOL, TEWZOK) with imine derivatives of 2-acetylpyridine, one of which was investigated as an antitumor agent [50]. The other example consists of two structures of Ag(I) (Ref. codes LAPCIR and YIMCAZ) and one structure of Cu(II) (Ref. code NEMCIR) in which one Schiff base moiety has the role of bidentate, while the other one acts as a tridentate ligand [51–53].

In two complexes – one complex of Cu(II) (Ref. code BABFET) and one of In(III) (Ref. code NAQFEQ) one Schiff base ligand is coordinated, while the other one has a role of counterion, with protonated pyridine nitrogen atom [54, 55].

Besides, there are four complexes in which Schiff base has exclusively the role of the counter ion—one complex of thorium(IV) (Ref. code BEQVAU) [56] and one complex of uranium(VI) (Ref. code MOYRIB) [57] with nitrates as ligands, and one complex of zinc(II) (Ref. code QELKOK) and one complex of cadmium(II) (Ref. code QELLAX) with chlorides as ligands [58]. Here, the synthetic routes should be mentioned, that is, in the presence of the deprotonating agent the coordination of the Schiff base will take place, whereas in the absence of an agent for deprotonation the Schiff base will be present in its protonated form and have a role of a counterion.

Recently, a series of copper(II) complexes with 2-acetylpyridine-aminoguanidine have been synthesized and structurally characterized and their interactions with the chemotherapeutical drug, cisplatin, were investigated [1, 59].

3.3 2-acetylpyridine-aminoguanidine as ligand

Synthesis of the titled Schiff base was described back in 1999 [60]. The reaction of aminoguanidine hydrogen carbonate and 2-acetylpyridine in acidic medium was refluxed to give microcrystals of the Schiff base. Recently, due to slight modification of the synthetic procedure, the chloride and sulfate salts (**Figure 9**) of this ligand were obtained in the form of white prismatic single crystals [59]. Besides, the thiocyanate ligand salt was obtained by the reaction of chloride ligand salt with ammonium-thiocyanate [61]. In this double-protonated form of the Schiff base, both pyridine and imino nitrogen atoms of the aminoguanidine residue are protonated. These results enabled detailed structural comparison of the free ligand in its cationic form (**Figure 9**) and its coordinated forms.

Investigation of the coordination behavior of 2-acetylpyridine-aminoguanidine revealed that, due to the pronounced alkalinity of aminoguanidine residue nitrogen atoms, spontaneous deprotonation of the ligand is unlikely, thus the role of Schiff base in the formed complexes (ligand or counterion) should be defined by the presence (or the absence) of deprotonating agents. In the reaction of $ZnCl_2$ or $CdCl_2$ with chloride ligand salt, in the presence of LiOAc complexes of the formula $[M(L)Cl_2]{\cdot}nH_2O$ (n = 0, 1) are formed, while the same reactants in the absence of LiOAc give complexes in which coordination of Schiff base did not take place, thus it is double protonated and has a role of counterion (**Figure 10**) [58].

Upon coordination, the Schiff base retains a high degree of planarity and the same *E*-configuration at C2−N3 bond as observed for the protonated ligand. Deprotonation of the AG residue leads to shortening of the C1−N1 bond and elongation of the C1−N2 bond, while coordination of the ligand causes changes of the angles in aminoguanidine residue which bring N1 and N5 donors closer.

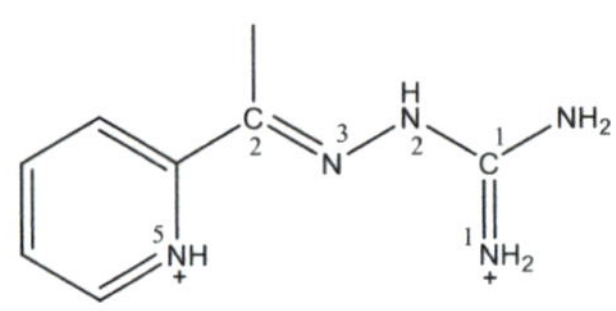

Figure 9.
Structural formula of the cation of 2-acetylpyridine-aminoguanidine.

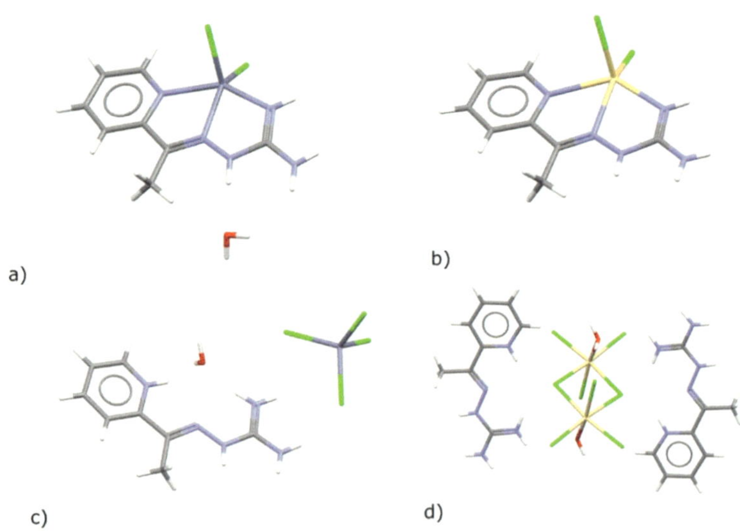

Figure 10.
Molecular structures of the complexes $[Zn(L)Cl_2]{\cdot}H_2O$ *(a),* $[Cd(L)Cl_2]$ *(b),* $[H_2L][ZnCl_4]{\cdot}H_2O$ *(c), and* $[H_2L]_2[Cd_2Cl_8(H_2O)_2]$ *(d).*

So far, with this ligand, five copper(II) complexes [59], three complexes of zinc(II) and cadmium(II) each [58, 61, 62], two complexes of iron(III) [63, 64], and one complex of cobalt(III) [63] were synthesized and structurally characterized.

In most of these complexes, Schiff base is coordinated in its neutral form, in the NNN tridentate manner, *via* pyridine, azomethine, and the nitrogen atom of the aminoguanidine residue. Monodeprotonated form of this ligand is realized by the deprotonation of the hydrazine nitrogen atom, and this form is achieved only in two iron(III) complexes.

In four complexes Schiff base is fully protonated and acts as a counterion, while there is only one complex in which monocation of the Schiff base is coordinated in a bidentate manner, while another cation of the Schiff base is neutralizing the charge of the complex anion. This is an interesting structure as the Cd(II) is situated in distorted octahedral environments, with two axial sites occupied by NCS^- ions, two equatorial sites occupied by bidentate coordinated organic ligand, and the remaining two equatorial sites (X, and Y) occupied by Cl^-/Br^- and Br^-/SCN^- ions, respectively. This means that the crystal structure contains four different anions: $[Cd(HL)(NCS)_2Br(SCN)]^-$ (ca. 61%), $[Cd(HL)(NCS)_2Cl(SCN)]^-$ (ca. 35%), $[Cd(HL)(NCS)_2Br_2]^-$ (ca. 3%), and $[Cd(HL)(NCS)_2ClBr]^-$ (ca. 1%).

4. Schiff bases of 2,6-diacetylpyridine

In this part, we will present the most common and interesting coordination modes, structural properties, and arrangement of Schiff bases of 2,6-diacetylpyridine (DAP) in their coordination complexes. The disubstituted pyridines show different properties, such as antimicrobial [65], anti-inflammatory, antitumor, analgetic [66], etc. activities. Some pyridine derivatives were used to prepare sensor membranes [67, 68], catalysts [69, 70], and ionophores for metal ion catching. Between them, a very important class is the 2,6-disubstituted pyridines, which are generally prepared by deriving DAP. The development of the one-pot-synthesis of DAP contributes to increase of its significance as a precursor for further syntheses [71]. Symmetrical disubstituted derivatives of DAP were obtained with different potential applicability, but mostly with significant biological activity [72]. The very important classes of DAPs are their thiosemicarbazones, hydrazones, and macrocyclic derivatives with promising biological activities. From the structural point of view, the symmetrically substituted DAPs can act as tri-, tetra-, or pentadentate ligands. On the other hand, the asymmetrical DAP derivatives characterize a smaller number of coordinative bonds (**Figure 11**). Both symmetrical and asymmetrical DAPs may coordinate as bridging or chelating ligands.

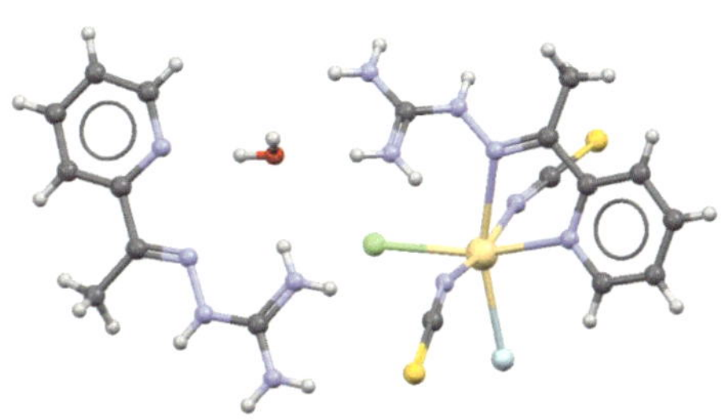

Figure 11.
Molecular structure of $[HL][Cd(HL)(NCS)_2XY]\cdot H_2O$, X = Cl^-/Br^- (teal ball), and Y = Br^-/SCN^- (green ball).

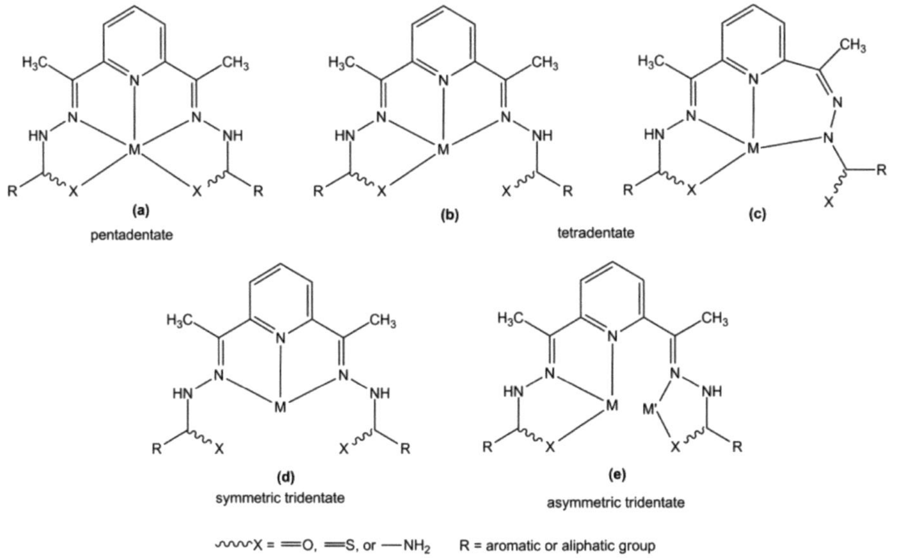

Figure 12.
Coordination modes of hydrazone-type Schiff bases of DAP - pentadentate (a), tetradentate (b and c), symmetric tridentate (d) and asymmetric tridentate (e).

The first is more characteristic for asymmetrical, while the last one is for symmetrical DAPs. The most common coordination modes of symmetric DAP-type Schiff bases are presented in **Figure 12**. Since DAP thiosemicarbazones are the focus of numerous studies [73–76], they would be described in the first subclass of DAPs.

Thiosemicarbazide (TSC) is known as a chemical precursor/intermediate during syntheses, reagent for ketones, and detection of certain metals [77–79]. Based on literature data, it was used in photography and as a rodenticide, herbicide, etc. [80]. The thiosemicarbazones of DAP (DAP-TSCs) may be synthesized easily by direct method or in the presence of a metal template. Numerous examples of DAP-TSCs with *d*-metals are given in the papers [76, 81–84]. In [76] a dimeric compound is presented. The ligand is a common DAP-TSC without additional substituents, formed from DAP and two TSCs in the presence of $LaCl_3$. In the dimeric complex, described in [76], DAP-TSC forms five coplanar coordinative bonds with zinc(II). Two additional axial coordination bonds are also determined with bromide or hydroxide, and with the sulfur of the nearby complex unit. The sulfur atom from one TSC fragment from each DAP-TSC ligand coordinates to both metal centers of the dimer forming a bridge between them. The central atoms and the sulfur bridges are in the same plane, which is not completely normal (85.61°) for the plane of the DAP-TSC ligand [76]. In another paper a complex of Pt(II) with 2,6-diacetylpyridine bis(^{4}N- tolylthiosemicarbazone) (DAP-TTSC) is described [81]. Before the complexation reaction, DAP-TTSC was prepared by direct synthesis. In the coordination compound, the Pt(II) is placed in the center of a square planar coordination polyhedron. The ligand's donor atoms are the hydrazine N, the pyridine, and the azomethine N atoms. The fourth position in the square planar arrangement is occupied by the S atom of the TTSC arm. The free ligand is symmetric as can be expected. During the reaction with K_2PtCl_4, one of the two tolyl rings is reduced and became cyclohexyl [81]. In the continuation of this study, Pt(II) Pd(II) compounds with 2,6-diacetylpyridine bis(^{4}N-tolylthiosemicarbazones) were characterized and tested against several cancer cell lines [84]. The metal ions, Pt(II) and Pd(II) in these compounds are in a square-planar environment and the *p*-tolyl substituted DAP-TTSC ligand coordinates to them asymmetrically. Namely, besides the pyridine N atom, donor atoms are the azomethine N and the S atom from the same TTSC arm. The other TSC arm coordinates to the metal center only through the azomethine N atom. This way of coordination results in thiol formation on the second TTSC arm. The DAP-TSC type ligands usually coordinate through pyridine and azomethine N atoms [65, 76, 85–89]. The coordination through hydrazine N is less common but also found, mostly in asymmetric complex compounds [81, 82, 84]. Due to the generally good biological activity of thiosemicarbazones, several papers address syntheses, spectral, and magnetic characterization, and tests of biological activity. Usually, antimicrobial, antifungal, or antiproliferative activity is tested on different cell lines. Such an example is [90] in which the procedure of synthesis of 2,6-diacetylpyridine bis(thiocarbohydrazone), 4-amino substituted thiocarbazone, and its reaction with Mn(II), Co(II) and Cu(II) nitrates, and acetates is described. The compounds were characterized by spectroscopic techniques [90]. It is supposed that the described compounds coordinate in a pentadentate manner and gives a positively charged complex unit. The antifungal and antibacterial tests have proved that the complex compounds are more active than the free ligand but less active than the reference substances against fungi and bacteria. Between them, the highest antifungal and antibacterial impact were shown by Cu(II) compounds.

The slightly modified thiosemicarbazides with DAP give symmetric bis(thiosemicarbazone) 2,6-diacetylpyridines under mild conditions, as described in [91]. They synthesized three monoligand coordination complexes of indium(III) with pentadentate DAP-TSCs and one binuclear compound. The ligands were prepared by direct synthesis from mixtures of DAP and thiosemicarbazide (TSC); 4-methyl-3-thiosemicarbazide (MTSC); 4,4-dimethyl-3-thiosemicarbazide (DTSC); and 4-phenyl-3-thiosemicarbazide (PTSC). The first three thiosemicarbazone-type ligands coordinate as pentadentate with the SNNNS donor set. The coordination is realized through the S atom of thiosemicarbazide fragments, the azomethine N, and the N atom from the pyridine ring. The ligands even during coordination kept their planar structure and the In(III) ion in the obtained coordination complexes with DAP-TSC, DAP-MTSC, and DAP-DTSC is placed in the center of a pentagonal bipyramidal coordination polyhedron. The equatorial plane is occupied by the DAP-TSC moiety, while the axial positions are saturated by chlorides from the starting $InCl_3$. Differently, the DAP-PTSC coordinates as a hexadentate. Namely, it is bonded as a pentadentate to the central In(III) ion like in the previous compounds, but a sulfur atom of one of the two thiosemicarbazone moieties from the ligand coordinates to the neighbor indium(III) ion, too, and *vice versa*. This In-S-In-S plane is analog to Zn-S-Zn-S moiety presented in [76]. The DAP moiety of the ligand occupies the equatorial plane of the pentagonal bipyramidal coordination polyhedron. However, this plane is more distorted than in the monomeric complexes. The bridging S atom occupies one of the two axial positions of the next In(III) and only one axially bonded chloride is involved in the complex formation (**Figure 13**).

The bridging sulfur atoms with the In(III) ions define a plane that is almost normal to the equatorial planes of the compound. The angle between the two planes is 85.50°. The lipophilic character of the thiosemicarbazone fragments of the ligand determines its anticancer activity. Compounds with higher lipophilicity are more active (more toxic) against cancer cells [91]. Despite the voluminous 4-phenyl substituted thiosemicarbazides and dimeric structure of its complex (**Figure 13**), the most lipophilic compound is the complex of DAP-DTSC. Another group of

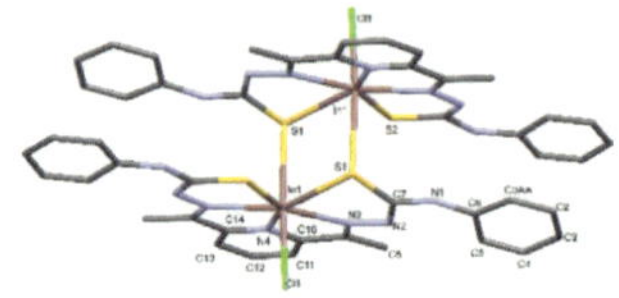

Figure 13.
The structure of the dimeric complex with the In-S-In-S plane.

authors describes a structurally very intriguing Bi(III) compound with promising anticancer and antimicrobial properties [87]. For ligand preparation besides DAP, [4]N-methylthiosemicarbazide was used and 2,6-diacetylpyridine bis(4-methylthiosemicarbazone) (DAP-MTSC) was obtained. This ligand reacts with $Bi(NO_3)_3$ under reflux for 1 h. The result of the reaction is a nine-coordinated Bi(III) compound. The DAP-MTSC ligand coordinates similarly to the complexes of indium (III) [91] in a pentadentate manner through its N_3S_2 donor set. The relatively rigid ligand has a distorted planar geometry after the coordination to Bi(III). The planarity is disturbed by the repulsion between the two sulfur atoms, which are placed slightly above and below the ligands plane. The probably most interesting detail in this structure is the bidentate coordination of two nitrates. Despite the weak coordination ability, in this complex two nitrates give in summary four Bi—O coordination bonds. The Bi—N and Bi—S bonds are not equal in length as was expected. Due to the different bond lengths, the coordination polyhedron around Bi(III) is distorted with the largest distance between the two S donor atoms. The thiosemicarbazone-type ligand coordinates through azomethine-, and pyridine N atoms as well as by sulfur atoms.

Due to the physiological significance and promising anticancer activity of Zn compounds and the significant biological activity of thiosemicarbazone-type Schiff bases, a series of Zn complexes with thiosemicarbazone-type ligands is synthesized to get a more potent multi-targeted anticancer drug [92]. The synthesis, structure, and mechanism of the anticancer effect of five Zn complexes with thiosemicarbazone moieties substituted differently in the [4]N position are given. Only one of the presented five complexes is monomeric with the 2,6-diacetylpyridine bis(4-phenylthiosemicarbazone) (DAP-PTSC) ligand. In this compound, the Schiff base ligand is coordinated as a pentadentate in a monodeprotonated form. The sixth coordination position is occupied by chloride. For Zn(II) ions the tetrahedral coordination geometry is the most favorable, but an octahedral arrangement is also acceptable. Since the ligand is pentadentate and relatively rigid, it cannot give a regular coordination polyhedron. Therefore, the compound presented in [92] is very distorted and fits neither in a tetragonal bipyramid (octahedron) nor in a pentagonal pyramid. The other four compounds with DAP-TSC type ligands prepared from thiosemicarbazide (TSC), N,N-diethylthiosemicarbazide (ETSC), N-methylthiosemicarbazide (MTSC), and N-tert(butyl)thiosemicarbazide (BTSC), are dimeric. In all complexes, one of the two thiosemicarbazone moieties of the ligand molecule is coordinated to one Zn(II), while the other moiety is coordinated to the other Zn(II) and *vice versa*. In the compound with DAP-TSC pyridine N atoms from both ligands are coordinated to the same Zn(II) ion, so that is hexacoordinated. To the other Zn(II) center only the TSC fragments in a tetrahedral arrangement are coordinated. The compound of DAP-ETSC is symmetrical and both Zn(II) centers are in a distorted trigonal bipyramidal environment with a coordinated NNS donor set from one of the DAP-ETSC ligands and an NS set from another one. The same coordination characteristics are observed in the case of the

DAP-BTSC complex. DAP-MTSC ligand with Zn(II) gives two different molecules per asymmetric unit. Namely, in one of these two molecules both Zn(II) are in the same distorted trigonal bipyramidal environment with NNS and NS donor sets from the two ligand molecules. In the other dimeric unit, one of the zinc(II) ions is coordinated in a distorted octahedral manner. To this metal center, N atoms of both ligands are bonded. The other Zn center is surrounded by tetrahedrally arranged NS donors of the ligand molecules. However, the focus of the paper [92] is on the mechanism of its anticancer activity, but the examined compounds are interesting from a structural aspect also. An explicit relationship between structure and anticancer activity is also found. The dimeric complexes with two Zn centers per complex molecule are more active against tumor cells than the monomeric compound. Another example of thiosemicarbazone type Schiff base is given in the paper [74]. This compound expresses good antioxidant activity, and additionally, easily reacts with transition metal ions, such as Ni(II) and Cu(II). The spectral studies have shown that both Cu(II) and Ni(II) are in an octahedral environment. From these data it may be concluded that the Schiff base ligand is coordinated as a tetradentate and in the axial positions are the chlorides from the starting metal salts. The complexes unfortunately show only minor antioxidant activity.

In addition, some compounds with S-substituted thiosemicarbazones are also published. Due to good coordination abilities and tunable biological activity, the S-methylisothiosemicarbazides (ITSC) have been a longer in focus of coordination chemistry studies. They can easily condense with carbonyl groups and give Schiff bases. Such a condensate is 2,6-diacetylpyridine bis(S-methylisothiosemicarbazone) (DAP-ITSC). With this ligand syntheses and physicochemical characterization of Mn(II), Co(II), Ni(II), and Zn(II) compounds are described [93]. The composition and most probable coordination of the ligand is supposed. Based on elemental analysis, conductometric data, and IR spectra one can conclude that the metal templated synthesis of the ligand was successful and resulted in complex formation. The IR spectra of Mn(II), Co(II), and Zn(II) complexes suggest the pentadentate coordination of the ligand through the 5N donor set. The two axial positions are most probably occupied by iodide and/or methanol and the final geometry is pentagonal bipyramidal. Differently, the collected data suggest that the Ni(II) complexes are cationic with pentadentate DAP-ITSC ligand, which is arranged in a helical trigonal bipyramidal geometry around Ni(II) ion. Neither the I^- nor the NCS^- ion is coordinated to Ni(II). These anions are counter ions in the outer coordination sphere. In the paper [94], a Ni(II) compound with the previous ligand is described. In the presented complex the DAP-ITSC ligand coordinates asymmetrically through the pyridine N; amino N and azomethine N from one ITSC arm, and hydrazine N atom from the other ITSC arm (**Figure 12c**). In this way, 4 N tetradentate coordination is achieved. Besides, an iodide is coordinated to Ni(II) in the apical position. The S-methyl substituents disturb the planarity of DAP-ITSC and the overall geometry of the compound is distorted square-pyramidal. An analogous compound with pyridine-2,6-dicarbaldehide and ITSC was synthesized and described in [95]. The dicarbaladehide-type ligand coordinates like the analog DAP derivative. One of the ITSC arms is deprotonated and coordinated as a bidentate to Ni(II), while the amino group of the other one is protonated and this one is coordinated as a monodentate. Together with the pyridine N, the ligand is coordinated by the 4 N donor set. By comparing the planarity of the same ligand in two complexes with Ni(II) [94] and Fe(III) [96] it is clear that the same DAP-ITSC ligand with different metal ions and in the presence of different anions is arranged differently in the coordination polyhedron.

Besides the thiosemicarbazones of 2,6-diacetylpyridine different carbazones and hydrazones of DAP are also significant in structural and antimicrobial aspects. Hydrazones are usually obtained by condensation of different hydrazines with DAP. The condensation reaction is relatively simple and can be carried out easily. By the procedure described in [97], three 2,6-diacetylpyridine bis(benzoylhydrazones) (DAP-BH) with R = H, Cl, and NO_2 substituents in the *para* position on the ring of the benzoyl moiety were prepared [98]. The authors have found that coordination complexes of such DAP-BH Schiff bases with Sb(III) and Bi(III) show activity against selected bacteria and tumor cell lines. The bismuth(III) compounds are more potent antibacterial and anticancer agents than the Sb(III) analogs. The DAP-BH type ligand coordinates symmetrical as a pentadentate through its ONNNO donor set. The [2,6-diacetylpyridine bis(p-chlorbenzoylhydrazone)]chlorido-antimony(III) has a distorted pentagonal pyramidal structure. Interestingly the otherwise planar DAP-BH in the basal plane of this complex bends toward the chloride apex of the pyramid, like an umbrella. The authors explained it with the lone electron pair of Sb(III) [98]. The Bi(III) complex beside the DAP-BH ligand and chloride contains a coordinated dimethyl sulfoxide (DMSO) in the other apex of the pentagonal bipyramid. The coordination geometry of this compound is also distorted like that in Sb(III) complex. Compounds of Ag(I) with the previous DAP-BHs, which express promising antifungal activity, were prepared and investigated also [99]. The pentadentate DAP-BH type Schiff bases can coordinate with lanthanides, such as Gd(III), Tb(III), Dy(III), and Er(III), too, and give stable single molecule- and single chain magnets [100–102], which keep their magnetism even after removal of the external magnetic field. The cobalt(II) compounds of 2,6-diacetylpyridine bis(4-hydroxybenzoylhydrazone) have an analogous structure to the previous DAP-BH complexes. It is coordinated as a planar pentadentate ligand. In the reference [103], the structure of a [Co(DAP-BH)$(NSC)_2$] is presented together with significant DNA binding and cleavage activity. The condensation of DAP with benzoyl or isonicotinoyl hydrazides can be carried out by direct synthesis. The condensation and hydrazone formation is also successful in the presence of Ni(II) ion. This metal-templated reaction gives positively charged complex ions [104]. These hydrazones depending on the extent of deprotonation can form mononuclear complexes in which they act as tri-, tetra-, or pentadentate ligands. They also can be at the same time chelating and bridging ligands that form binuclear dimeric coordination compounds [104]. The nicotinoyl- and isonicotinoyl hydrazone (DAP-NH and DAP-INH) analogs of previous benzoylhydrazone type Schiff bases are presented in [105]. The coordination of DAP-INH is analogous to the previous hydrazone type ligand with the ONNNO donor set. The apices of the coordination polyhedron are occupied by ethyl groups. This coordination complex is cationic with a 2+ charge. The anions necessary for electroneutrality are chlorides and a diethyltrichloridostanate(II) ion. Recently 2D coordination polymers were prepared from the same DAP-NH and DAP-INH ligands and Mn(II). The final composition as well as the discrete or polymeric structure of the compounds depends on reaction conditions [106]. By template synthesis in the presence of KNCS both $MnCl_2$ and $Mn(NO_3)_2$, give discrete coordination complexes. Differently, by direct solvothermal synthesis 2D polymers were obtained. In these polymeric chains, the DAP-NH and DAP-INH ligands are coordinated pentadentately as previously (in one plane), and the N atoms of the neighbor nicotinoyl/isonicotinoyl moieties coordinate to Mn(II) below and above the DAP-NH/DAP-INH plane, apically. By solvothermal method of synthesis, homo- and heterometallic coordination polymers with DAP-NH Schiff base were obtained and described [107]. The 2D coordination polymers presented in

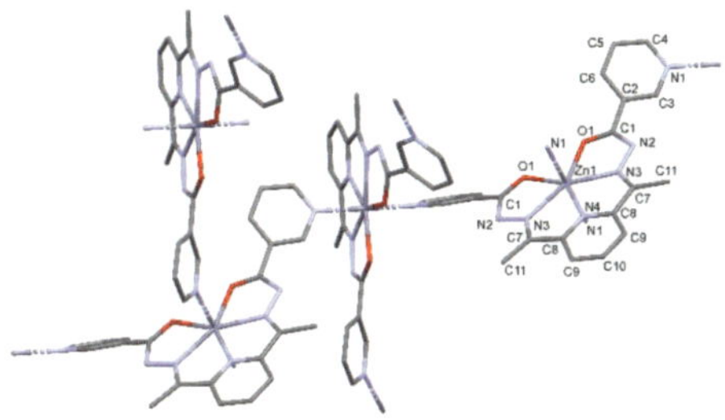

Figure 14.
Coordination modes of DAP-NH Schiff base in a 2D coordination polymer.

Ref. [107] due to the free space in their crystal lattices may adsorb solvent molecules and accommodate them until relatively high temperatures. In these polymers, the DAP-NH ligand is coordinated as a pentadentate for one metal center and at the same time is coordinated through one or both of its nicotinoic N atoms to the neighbor metal ion/ions. The pentadentate DAP-hydrazone moiety gives the equatorial plane of the pentagonal bipyramid coordination polyhedron around the metal center. The nicotinoyl rings coordinate in the apical positions of the neighbor metal ions, forming chains in such a way (**Figure 14**).

A characteristic hydrazide that easily can react with DAP is trimethylammonium acetohydrazide, known as Girard's T reagent. The preparation procedure of 2,6-diacetylpyridine bis(trimehylammoniumacetohydrazone) (DAP-GT) at first was published in a study by Main and Fritz [108]. Later this procedure was improved, and the same Schiff base somewhat easier and faster was got [109]. The structure of the coordination complex of DAP-GT with Mn(II) is pentagonal bipyramidal as the other 2,6-diacetylpyridine bis(hydrazones) with ONNNO donor set in the equatorial plane and the N ends of thiocyanates coordinated in axial positions. Beside Mn(II), with the same ligands were prepared coordination complexes of Co(II), Ni(II) [88], Zn(II), Cd(II) [110], and Fe(III) [111]. A group of 2,6-diacetylpyridine bis(benzenesulfonohydrazones) (DAP-BS) was synthesized by reflux of acidic reaction mixtures of DAP with benzenesulfonohydrazide (BS), 4-methylbenzenesulfonohydrazide (MBS), and 4- methoxybenzenesulfonohydrazide (OBS) [112]. The Cu(II) complexes of the DAP-BSs show significant inhibition of Methicillin-resistant strains of *Staphylococcus aureus*. From the prepared Schiff bases and copper(II) complexes, only one coordination complex was precipitated as a single crystal, with 2,6-diacetylpyridine bis(4-methoxybenzenesulfonohydrazone) (DAP-OBS). In this compound, the Schiff base ligand is coordinated as a tetradentate with an NNNO donor set (**Figure 12c**). Besides the pyridine and two azomethine N atoms, an O from one sulfonyl group is the fourth coordinating atom. Additionally, one pyridine is also coordinated and a distorted tetragonal pyramidal geometry is formed around Cu(II). Because of the geometry of sulfonyl groups, the benzene rings are displaced from the basal plane of the coordination polyhedron. Interestingly, the apical pyridine ligand and the ring of the coordinated benzenesulfonyl moiety are oriented similarly, very close to being parallel. In another paper of the same group, the molecular structure of unsubstituted DAP-BS and its Zn complex is determined [113]. In the Zn(II) compound, the Schiff base ligand is coordinated as a pentadentate through pyridine and two azomethine N atoms and oxygens from both benzenesulfonyl groups. Additionally, two DMSO molecules are coordinated to Zn(II) apically. The ligands together form a pentagonal bipyramidal surrounding of Zn(II). Recently the crystal structure of 2,6-diacetylpyridine bis(phenylhydrazone) (DAP-PH) and its mixed-valence copper complex is published [114]. In the described compound two DAP-PH Schiff bases coordinate to Cu(II) and give a bis(ligand) complex cation. DAP-PH coordinates as a tridentate with the NNN donor set. The counter ion consists of two Cu(I) bridged by two bromides and with a terminal bromide bonded to each Cu(I).

An interesting group of DAP hydrazones is prepared from biologically active compounds. Copper(II) complex of condensed 2,3-diamino-1,4-naphthoquinone with DAP shows inhibition of *S. aureus* comparable with inhibition of penicillin [115]. With glucosamine and ethanolamine, DAP also may give adducts. The reaction is Fe(III)-templated [116]. The authors give data about the antibacterial and radical scavenging activity of the compounds. The crystal structures of transition metal complexes with the antihypertensive phtalazine-1-hydrazones, also known as hydralazine (HZ) were published in [117–119]. It was found that the Schiff base 2,6-diacetylpyridine bis(phtalazine-1-hydrazone) (DAP-HZ) is a good ligand for the formation of binuclear complexes [117]. Nickel(II) gives a binuclear compound with DAP-HZ in which the Schiff base DAP-HZ acts as a hexadentate bridging ligand. Namely, one of the arms of DAP-HZ coordinates through pyridine N, azomethine N, and N atom from the hydralazine moiety to the first Ni(II) center, while the other arm through the analogous atoms to the second Ni(II) center. The pyridine N atom is coordinated with both Ni(II) ions. The other DAP-HZ coordinates with these two Ni(II) in the same way and forms a binuclear coordination compound with octahedral coordination geometry around both nuclei (**Figure 15**).

About 20 years later the structures, detailed characterization, and antimicrobial activity of mononuclear transition metal (Co(III), Ni(II), and Cu(II)) complexes of DAP-HZ were described in Ref. [119]. In the discussed compounds DAP-HZ acts as a monodeprotonated tetradentate ligand with a 4 N donor set. It gives the equatorial or basal plane of the complexes. One of the hydralazine moieties is not coordinated and bends out of the plane of the rest of the molecule. The electroneutrality of the compounds is achieved with chlorides. Accordingly, Co(III) are bonded two chlorides in the axial positions, and an octahedral complex is formed. The square planar geometry, characteristic of Ni(II), is realized by coordination only of DAP-HZ. Since the Schiff base is 1− charged, the other necessary negative charge ensured the chloride as a counter ion. The Cu(II) is in a tetragonal pyramidal environment with the chloride coordinated to the Cu(II) center in the apical position. In the reaction with Zn(II), DAP-HZ is additionally deprotonated and as a bianionic ligand gives a neutral complex. The influence of these compounds and some other aromatic Schiff

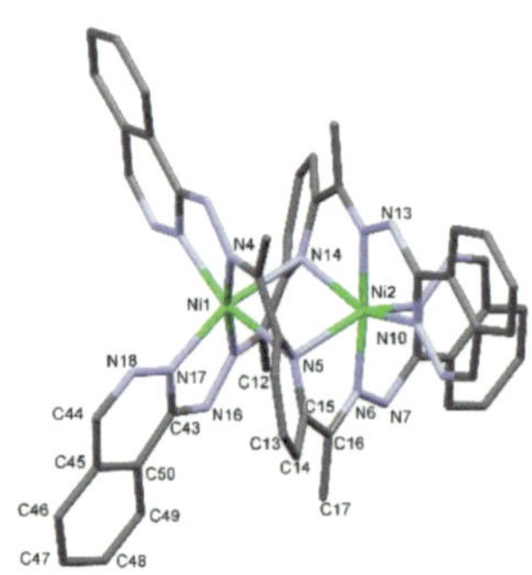

Figure 15.
Coordination mode of DAP-HZ in the binuclear Ni(II) compound.

bases on the anticancer activity of cisplatin is described in [1]. Another biologically active aromatic compound with a functional group analogous to HZ is 3-chloro-6-hydrazinopyridazine (HP). The corresponding pyridazone analog of DAP was prepared by direct synthesis. The coordination ability of obtained 2,6-diacetylpyridine bis(3-chloropyridazine-6-hydrazone) (DAP-HP) was investigated in reaction with Co(II), Ni(II), Cu(II), and Zn(II) [118]. DAP-HP deprotonates during complex formation and gives neutral coordination complexes with the metal ions. In the reaction of cobalt(II) acetate with DAP-HP, in the presence of methanol, Co(II) is oxidized to Co(III) and a binuclear compound is formed. One of the arms of DAP-HP coordinates as a tridentate to one Co(III) center through pyridine, azomethine, and pyridazine N atoms, while to the other Co(III) center as a bidentate through azomethine and pyridazine N atoms of the other arm. The other DAP-HP ligand coordinates in the opposite way. The sixth coordination position of both Co(III) centers is occupied by a peroxide bridge. Based on IR, Raman, molar conductivity, and thermal data, the Ni(II), Cu(II), and Zn(II) complexes are mononuclear with a square-planar arrangement of the tetradentate DAP-HP ligand.

Besides different carbazones and hydrazones, DAP can give Schiff base-type macrocyclic compounds. Some macrocyclic compounds show significant biological activity. Such compounds were described in [66]. The obtained semicarbazide pyridine derivative showed slight anticonvulsant and analgetic activity. Differently, 2,6-bis-[β-(2-thienyl)acryloylphenylsemicarbazide]-pyridine derivative showed activity comparable with referent drugs [66]. The macrocyclic compounds may also act as ligands for coordination compounds. The template synthesis of Mn(II), Zn(II), and Cd(II) complexes of a DAP derivative macrocyclic ligand is given in [120]. The Zn(II) compound is active against some bacteria and cell lines, while the antioxidant activity of the Mn(II) compound is similar to that of the reference substance. The macrocyclic ligands DAP moiety coordinates to the metal center similarly to the earlier compounds as a tridentate. The piperazine moiety forms two additional coordination bonds with the central atom through N atoms of the piperazine ring.

5. Conclusion

The significance, syntheses, structural properties, and potential or real application of the dihydroxybenzaldehide, 2-acetylpyridine, and 2,6-diacetylpyridine Schiff bases are described. Besides, for the first two groups of compounds, the search for CSD was made and the number and the frequency of certain coordination modes were investigated. Some interesting examples with mostly biological activity are collected with an accent on their structure and the coordination modes of DAP in them. There are many more Schiff bases of these compounds, than those presented in this summary with different potential or real applications. This shows the importance of future research in the field of Schiff bases and their metal complexes.

Acknowledgements

The authors acknowledge the financial support of the Ministry of Education, Science, and Technological Development of the Republic of Serbia (Grant No. 451-03-9/2021-2114/200125) and the Provincial Secretariat for Higher Education and Scientific Research (Grant No. 42-451-2197/2022).

Author details

Mirjana M. Radanović* and Berta Barta Holló
University of Novi Sad Faculty of Sciences, Novi Sad, Serbia

*Address all correspondence to: mirjana.lalovic@dh.uns.ac.rs

References

[1] Magyari J, Barta Holló B, Radanović MM, Armaković S, Armaković SJ, Molnár J, et al. Interactions of Schiff base compounds and their coordination complexes with the drug cisplatin. New Journal of Chemistry. 2018;**42**:5834-5843

[2] Gwaram NS, Ali HM, Khaledi H, Abdulla MA, Hadi HA, Lin TK, et al. Antibacterial evaluation of some Schiff bases derived from 2-acetylpyridine and their metal complexes. Molecules. 2012;**17**(5):5952-5971

[3] da Silva CM, da Silva DL, Modolo LV, Alves RB, de Resende MA, CVB M, et al. Schiff bases: A short review of their antimicrobial activities. Journal of Advanced Research. 2011;**2**(1):1-8

[4] Raczuk E, Dmochowska B, Samaszko-Fiertek J, Madaj J. Different Schiff Bases-Structure, Importance and Classification. Molecules. 2022;**27**(3):787

[5] Barajas-Carrillo VW, Estolano-Cobián A, Díaz-Rubio L, Ayllón-Gutiérrez RR, Salazar-Aranda R, Díaz-Molina R, et al. Antioxidant and acetylcholinesterase inhibition activity of aliphatic and aromatic edaravone derivatives. Medicinal Chemistry Research [Internet]. 2021;**30**(3):610-623. https://link.springer.com/article/10.1007/s00044-020-02667-5

[6] Groom CR, Bruno IJ, Lightfoot MP, Ward SC. The Cambridge structural database. Acta Crystallographica Section B: Structural Science, Crystal Engineering and Materials. 2016;**72**(2):171-179

[7] Khandar AA, Hosseini-Yazdi SA, Zarei SA. Synthesis, characterization and X-ray crystal structures of copper(II) and nickel(II) complexes with potentially hexadentate Schiff base ligands. Inorganica Chimica Acta. 2005;**358**(11):3211-3217

[8] Cai YP, Su CY, Xu A-W, Kang BS, Tong YX, Liu HQ, et al. Syntheses and characterization of copper(II) complexes of bis(acetylacetone)trimethylenediimine. Polyhedron. 2001;**20**(7-8):657-662

[9] Hökelek T, Kilic Z, Isiklan M, Toy M. Intramolecular hydrogen bonding and tautomerism in Schiff bases. Part II. Structures of 1-[N-(2-pyridyl) aminomethylidene]-2(1H)-naphtalenone (1) and bis[2-hydroxy-κO–N-(2-pyridyl)-1-naphthaldiminato-κN] zinc(II) (2). Journal of Molecular Structure. 2000;**523**(1-3):61-69

[10] Karabiyik H, Ocak-Iskeleli N, Petek H, Albayrak Ç, Ağar E. An intermediate structure trapped in solid-state tautomerization process of (E)-4-[(4-chlorophenylimino)methyl] benzene-1,2,3-triol. Journal of Molecular Structure. 2008;**873**(1-3):130-136

[11] Dey DK, Dey SP, Elmali A, Elerman Y. Molecular structure and conformation of N-2-[3′-(methoxysalicylideneimino) benzyl]-3″-methoxysalicylideneimine. Journal of Molecular Structure. 2001;**562**(1-3):177-184

[12] Alpaslan G, MacIt M, Erdönmez A, Büyükgüngör O. Experimental and computational studies on zwitterionic (E)-1-((4-phenoxyphenyliminio) methyl)naphthalen-2-olate. Journal of Molecular Structure. 2011;**1-3**(997):70-77

[13] Bandyopadhyay D, Granados JC, Short JD, Banik BK. Polycyclic aromatic compounds as anticancer agents: Evaluation of synthesis and in vitro

cytotoxicity. Oncology Letters. 2012;**3**(1):45

[14] Khosravi I, Hosseini F, Khorshidifard M, Sahihi M, Rudbari HA. Synthesis, characterization, crystal structure and HSA binding of two new N,O,O-donor Schiff-base ligands derived from dihydroxybenzaldehyde and tert-butylamine. Journal of Molecular Structure. 2016;1119:373-384

[15] Zaugg RH, Walder JA, Klotz IM. Schiff base adducts of hemoglobin. Modifications that inhibit erythrocyte sickling. Journal of Biological Chemistry. 1977;**252**(23):8542-8548

[16] Das U, Pattanayak P, Patra D, Brandão P, Felix V, Chattopadhyay S. Design, synthesis and properties of orthopalladated complexes: Proheterogeneous catalyst. Polyhedron. 2016;**110**:165-171

[17] Bouhidel Z, Cherouana A, Durand P, Doudouh A, Morini F, Guillot B, et al. Synthesis, spectroscopic characterization, crystal structure, Hirshfeld surface analysis and antimicrobial activities of two triazole Schiff bases and their silver complexes. Inorganica Chimica Acta. 2018;**482**:34-47

[18] Chakraborty A, Goura J, Bag P, Bar AK, Sutter JP, Chandrasekhar V. Windmill-shaped octanuclear ZnII4/LnIII4 (LnIII = Dy, Tb, Ho) heterometallic ensembles supported by a tetraferrocene scaffold. Dalton Transactions. 2016;**45**(44):17633-17643

[19] Trujillo A, Fuentealba M, Carrillo D, Manzur C, Ledoux-Rak I, Hamon JR, et al. Synthesis, spectral, structural, second-order nonlinear optical properties and theoretical studies on new organometallic donor-acceptor substituted nickel(ii) and copper(ii) unsymmetrical schiff-base complexes. Inorganic Chemistry. 2010;**49**(6):2750-2764

[20] Kaluarachchige Don UI, Kurup SS, Hollingsworth TS, Ward CL, Lord RL, Groysman S. Synthesis and Cu(I)/Mo(VI) reactivity of a bifunctional Heterodinucleating ligand on a xanthene platform. Inorganic Chemistry. 2021;**60**(19):14655-14666

[21] Zaltariov MF, Vlad A, Cazacu M, Avadanei M, Vornicu N, Balan M, et al. Silicon-containing bis-azomethines: Synthesis, structural characterization, evaluation of the photophysical properties and biological activity. Spectrochimica Acta. Part A, Molecular and Biomolecular Spectroscopy. 2015;**138**:38-48

[22] Tarai A, Baruah JB. Inclusion of 2,4-dihydroxybenzaldehyde and 2,4-dihydroxybenzaldoxime in cadmium coordination polymer and conversion of guest aldehyde to oxime. ChemistrySelect. 2017;**2**(35):11482-11486

[23] Luo J, Ma Z, Liang H, Chen J, Zeng Z. K, Ca complexes with a sulfonic ligand: Structure and DNA-binding properties. Spectrochimica Acta. Part A, Molecular and Biomolecular Spectroscopy. 2012;**90**:202-207

[24] Belmonte MM, Escudero-Adán EC, Martin E, Kleij AW. Isolation and characterization of unusual multinuclear Schiff base complexes: Rearrangements reactions and octanuclear cluster formation. Dalton Transactions. 2012;**41**(17):5193-5200

[25] Tang BB, Ma H, Li GZ, Wang YB, Anwar G, Shi R, et al. Synthesis, crystal structures and luminescent properties of tetranuclear Zn molecular clusters with aroylhydrazone ligand. CrystEngComm. 2013;**15**(40):8069-8073

[26] Djouhra A, Ali O, Ramiro RR, Emilia M. A selective naked-eye chemosensor derived from 2-methoxybenzylamine and 2,3-dihydroxybenzaldehyde—synthesis, spectral characterization and electrochemistry of its bis-bidentates Schiff bases metal complexes. Spectrochimica Acta. Part A, Molecular and Biomolecular Spectroscopy. 2017;**184**:299-307

[27] Amiri Rudbari H, Khorshidifard M, Askari B, Habibi N, Bruno G. New asymmetric Schiff base ligand derived from allylamine and 2,3-dihydroxybenzaldehyde and its molybdenum(VI) complex: Synthesis, characterization, crystal structures, computational studies and antibacterial activity together with synergistic effect agai. Polyhedron. 2015;**100**:180-191

[28] Mahmoud MA, Zaitone SA, Ammar AM, Sallam SA. Synthesis, structure and antidiabetic activity of chromium(III) complexes of metformin Schiff-bases. Journal of Molecular Structure. 2016;**1108**:60-70

[29] Radanović MM, Rodić MV, Vojinović-Ješić LS, Kostić MS, Gazdić D, Leovac VM. Synthesis and structure of tetranuclear zinc complex with Schiff base of aminoguanidine. In: 27th Conference of the Serbian Crystallographic Society. Belgrade, Serbia: Serbian Crystallographic Society; 2021. pp. 82-83

[30] Watala C, Dobaczewski M, Kazmierczak P, Gebicki J, Nocun M, Zitnanova I, et al. Resorcylidene aminoguanidine induces antithrombotic action that is not dependent on its antiglycation activity. Vascular Pharmacology. 2009;**51**(4):275-283

[31] Onuska KD, Taylor NJ, Carsky J. Crystal and molecular structure of β-resorcylidene aminoguanidine copper(II) complex. Journal of Chemical Crystallography 1996 26:12. 1996;26(12):841-846

[32] Khatkar P, Asija S, Singh N. Synthesis, spectral studies and in vitro antimicrobial activity of some new Di/Triorganotin (IV) complexes of Schiff bases derived from 2-benzoylpyridine. Journal of the Serbian Chemical Society. 2017;**82**(1):13-23

[33] Salam MA, Affan MA, Ahmad FB, Arafath MA. Organotin(IV) complexes with pyruvic acid phenylhydrazone (HPAPD): Synthesis, spectral characterization, and in vitro antibacterial activity. Journal of Coordination Chemistry. 2012;**65**(11):1999-2007

[34] Tang B, Yu F, Li P, Tong L, Duan X, Xie T, et al. A near-infrared neutral pH fluorescent probe for monitoring minor pH changes: Imaging in living HepG2 and HL-7702 Cells. Journal of the American Chemical Society. 2009;**131**(8):3016-3023

[35] Song B, Wang G, Tan M, Yuan J. A europium(III) complex as an efficient singlet oxygen luminescence probe. Journal of the American Chemical Society. 2006;**128**(41):13442-13450

[36] Sánchez-Ponce L, Galindo-RiañoMD,Casanueva-MarencoMJ, Granado-Castro MD, Díaz-De-alba M. Sensing Cd(II) using a disposable optical sensor based on a Schiff Base immobilisation on a polymer-inclusion membrane. Applications in water and art paint samples. Polymers (Basel). 2021;**13**(24):4414

[37] Chavan SS, Sawant SK, Sawant VA, Lahiri GK. Mixed-ligand complexes of copper(I) with Schiff base and triphenylphosphine: Effective catalysts for the amination of aryl halide. Inorganic Chemistry Communications. 2011;**14**(9):1373-1376

[38] Miller MC, Stineman CN, Vance JR, West DX, Hall IH. The cytotoxicity of copper(II) complexes of 2-acetyl-pyridyl-4N-substituted thiosemicarbazones. Anticancer Research. 1998;**18**(6A):4131-4139

[39] Miller MC III, Stineman CN, Vance JR, West DX, Hall IH. Multiple mechanisms for cytotoxicity induced by copper(II) complexes of 2-acetylpyrazine-N-substituted thiosemicarbazones. Applied Organometallic Chemistry. 1999;**13**:9-19

[40] Garcia CV, Parrilha GL, Rodrigues BL, PJS B, Clarke RM, Storr T, et al. Cobalt(III) complexes with 2-acetylpyridine-derived Schiff bases: Studies investigating ligand release upon reduction. Polyhedron. 2017;**124**:86-95

[41] Ferraz KSO, da Silva JG, Costa FM, Mendes BM, Rodrigues BL, dos Santos RG, et al. N(4)-Tolyl-2-acetylpyridine thiosemicarbazones and their platinum(II,IV) and gold(III) complexes: cytotoxicity against human glioma cells and studies on the mode of action. Biometals. 2013;**26**(5):677-691

[42] Parrilha GL, Ferraz KSO, Lessa JA, de Oliveira KN, Rodrigues BL, Ramos JP, et al. Metal complexes with 2-acetylpyridine-N(4)-orthochlorophenylthiosemicarbazone: Cytotoxicity and effect on the enzymatic activity of thioredoxin reductase and glutathione reductase. European Journal of Medicinal Chemistry. 2014;**84**: 537-544

[43] Xie J, Shen S, Chen R, Xu J, Dong K, Huang J, et al. Synthesis, characterization and antitumor activity of ln(III) complexes with hydrazone schiff base derived from 2-acetylpyridine and isonicotinohydrazone. Oncology Letters. 2017;**13**(6):4413-4419

[44] Mohammed Hosny N, Sherif YE, El-Rahman AA. Spectral characterization and anti-inflammatory activity of Schiff-base complexes derived from leucine and 2-acetylpyridine. Journal of Coordination Chemistry. 2008;**61**(16):2536-2548

[45] Zhang N, Fan YH, Zhang Z, Zuo J, Zhang PF, Wang Q, et al. Syntheses, crystal structures and anticancer activities of three novel transition metal complexes with Schiff base derived from 2-acetylpyridine and l-tryptophan. Inorganic Chemistry Communications. 2012;**22**:68-72

[46] Deghadi RG, Elsharkawy AE, Ashmawy AM, Mohamed GG. Antibacterial and anticorrosion behavior of bioactive complexes of selected transition metal ions with new 2-acetylpyridine Schiff base. Applied Organometallic Chemistry. 2022;**36**(4):e6579

[47] Despaigne AAR, Parrilha GL, Izidoro JB, da Costa PR, dos Santos RG, Piro OE, et al. 2-acetylpyridine- and 2-benzoylpyridine-derived hydrazones and their gallium(III) complexes are highly cytotoxic to glioma cells. European Journal of Medicinal Chemistry. 2012;**50**:163-172

[48] Biswas C, Chatterjee A, Vijayan V, Purohit CS, Kiran MS, Ghosh R. Synthesis, structural characterization and selective anticancer activity of $[AgI(L)(PPh3)]_2(NO_3)_2$ [L = N(4)-substituted 2-acetylpyridine-N(4)-methyl-3-thiosemicarbazone]. Inorganic Chemistry Communications. 2022;**136**:109178

[49] Kovala-Demertzi D, Demertzis MA, Filiou E, Pantazaki AA, Yadav PN, Miller JR, et al. Platinum(II) and palladium(II) complexes with 2-acetyl pyridine 4N-ethyl thiosemicarbazone able to overcome the cis-platin resistance.

Structure, antibacterial activity and DNA strand breakage. Biometals. 2003;**16**(3):411-418

[50] Kovala-Demertzi D, Demertzis MA, Miller JR, Frampton CS, Jasinski JP, West DX. Structure of bis(2-acetylpyridine 3-hexamethylene iminylthiosemicarbazonato) palladium II, a potential antitumor complex. Journal of Inorganic Biochemistry. 2002;**92**(2):137-140

[51] Onodera K, Kasuga NC, Takashima T, Hara A, Amano A, Murakami H, et al. Synthesis, reaction and structure of a highly light-stable silver(I) cluster with an $Ag_4S_4N_4$ core having a tridentate 4N-morpholyl 2-acetylpyridine thiosemicarbazone ligand: Use of water-soluble silver(I) carboxylates as a silver(I) source. Dalton Transactions. 2007:3646-3652

[52] Huang C, Chen DM, Zhu BX. Syntheses and crystal structures of Hg(II) and Ag(I) complexes containing 2-acetylpyridine-paminobenzoylhydrazone ligand. Chinese Journal of Structural Chemistry. JIEGOU HUAXUE. 2017;**36**(3):471-477

[53] Bermejo E, Castiñeiras A, Ackerman LJ, Owens MD, West DX. Structural studies of a copper(II) complex with Inequivalent2-acetylpyridine 3-Piperidylthiosemicarbazonato ligands and a nickel(II) complex with different modes of coordination by the two Thiosemicarbazonato moieties of 1-phenylglyoxal bis{N(4)-diethylthiosemicarbazone}. Zeitschrift für Anorganische und Allgemeine Chemie. 2001;**627**:1966-1971

[54] Xue LW, Zhang HJ, Wang PP. Synthesis, crystal structures, and antimicrobial activity of copper(II) complexes derived from N′-(1-(pyridin-2-yl)ethylidene)isonicotinohydrazide. Inorganic and Nano-Metal Chemistry. 2020;**50**(8):637-643

[55] Abram S, Maichle-Mössmer C, Abram U. Synthesis and characterization of indium(III) complexes with tri- and pentadentate thiosemicarbazones. Crystal and molecular structure of $[InCl_2(HDAPTSC)] \cdot 2$ DMSO, {O[In(HDAPTSC)(OH)]2} · 5MeOH, $[InCl_2(APTSC)(MeOH)]$, [In(APTSC)2]PF6 and (H2APTSC) [InCl(APTSC)(mnt)] · 0.5 H_2O (H2DAPTSC = 2,6-diacetylpyridine-bis(thiosemicarbazone), HAPTSC = 2-acetylpyridine-thiosemicarbazone, mnt2 = 1,2-dicyanoethene-1,2-dithiolate). Polyhedron. 1998;**17**(1):131-143

[56] Abram U, Bonfada E, Schulz LE. Bis{2-[1-(thiosemicarbazono)ethyl] pyridinium} hexakis(nitrato-O,O') thorate(IV) tetramethanol solvate. Acta Crystallographica Section C. 1999;**55**(9):1479-1482

[57] Abram U, Schulz Lang E, Bonfada E. Thiosemicarbazone complexes of uranium. Zeitschrift für Anorganische und Allgemeine Chemie. 2002;**628**(8):1873-1878

[58] Radanović MM, Rodić MV, Vojinović-Ješić LS, Armaković S, Armaković SJ, Leovac VM. Complexes of Zn(II) and Cd(II) with 2-acetylpyridine - aminoguanidine—Syntheses, structures and DFT calculations. Inorganica Chimica Acta. 2018;**473**:160-168

[59] Vojinović-Ješić LS, Radanović MM, Rodić MV, Živković-Radovanović V, Jovanović LS, Leovac VM. Syntheses and characterization of 2-acetylpyridine-aminoguanidine and its copper(II) complexes: Crystallographic and antimicrobial study. Polyhedron. 2016;**117**:526-534

[60] Györgydeák Z, Holzer W, Mereiter K. Guanylhydrazones of (hetero)aryl methyl ketones: Structure and reaction with acetic anhydride. Monatshefte für Chemie/Chemical Monthly. 1999;**130**(7):899-913

[61] Radanović MM, Novaković SB, Vojinović-Ješić LS, Rodić M v, Leovac VM. 2-Acetylpyrydine-aminoguanidine—Novel ligand salt and zinc(II) complex containing thiocyanate. Journal of the Serbian chemical. Society. 2018;**83**(2):157-166

[62] Rodić M v, Radanović MM, Vojinović-Ješić L, Leovac V. Four-component isomorphism in the crystal structure of [HL][Cd(HL)$(NCS)_2$XY]·H_2O, where L = 2-acetylpyridine-aminoguanidine, X = Cl/Br and Y = Br/SCN. Acta Crystallographica Section A: Foundations and Advances. 2019;**75**(a2):e602-e602

[63] Vojinović-Ješić L, Radanović MM, Rodić M v, Leovac V. 2-acetylpyridine-aminoguanidine coordinating iron(III) and cobalt(III): Syntheses and characterization. In: IV International Congress "Engineering, Environment and Materials in Processing Industry. Faculty of Technology. University of East Sarajevo. Zvornik, Republic of Srpska, Bosnia and Herzegovina. 2017. pp. 1256-1266

[64] Radanović MM, Vojinović-Ješić LS, Rodić MV, Leovac VM. Synthesis and structure of iron(III) complex with 2-acetylpyridine-aminoguanidine. Contemporary Materials. 2021;**12**(1):64-70

[65] Filipović NR, Borrmann H, Todorović T, Borna M, Spasojević V, Sladić D, et al. Copper(II) complexes of N -heteroaromatic hydrazones: Synthesis, X-ray structure, magnetic behavior, and antibacterial activity. Inorganica Chimica Acta. 2009;**362**:1996-2000

[66] Amr AGE. Synthesis of some new linear and chiral macrocyclic pyridine carbazides as analgesic and anticonvulsant agents. Zeitschrift für Naturforschung B. 2005;**60b**:990-998

[67] Hassan SSM, Abou Ghalia MH, Amr AGE, Mohamed AHK. Novel thiocyanate-selective membrane sensors based on di-, tetra-, and hexa-imidepyridine ionophores. Analytica Chimica Acta. 2003;**482**(1):9-18

[68] Hassan SSM, Abou Ghalia MH, Amr AGE, Mohamed AHK. New lead(II) selective membrane potentiometric sensors based on chiral 2,6-bis-pyridinecarboximide derivatives. Talanta. 2003;**60**(1):81-91

[69] Fernandes S, Bellabarba RM, Ribeiro AF, Gomes PT, Ascenso JR, Mano JF, et al. Polymerisation of ethylene catalysed by mono-imine-2,6-diacetylpyridineiron/methylaluminoxane (MAO) catalyst system: Effect of the ligand on polymer microstructure. Polymer International. 2002;**51**:1301-1303

[70] Gupta KC, Sutar AK. Catalytic activities of Schiff base transition metal complexes. Coordination Chemistry Reviews. 2008;**252**(12-14):1420-1450

[71] Ivchenko P, v, Nifant'ev IE, Buslov I v. A convenient approach for the synthesis of 2,6-diformyl- and 2,6-diacetylpyridines. Tetrahedron Letters. 2013;**54**(3):217-219

[72] Anacona JR, Calvo G, Camus J. Tetradentate hydrazone metal complexes derived from cefazolin and 2,6-diacetylpyridine hydrazide: synthesis, characterization , and antibacterial activity. Monatshefte

für Chemie—Chemical Monthly. 2016;**147**(4):725-733

[73] Vieira RP, Thompson JR, Beraldo H, Storr T. Partial conversion of thioamide into nitrile in a copper(II) complex of 2,6-diacetylpyridine bis(thiosemicarbazone), a drug prototype for Alzheimer's disease. Acta Crystallographica Section C: Structural Chemistry. 2015;**71**:430-434

[74] Subba Rao Y, Prathima B, Adinarayana Reddy S, Madhavi K, Varada RA. Complexes of Cu(II) and Ni(II) with bis(phenylthiosemicarbazone):Synthesis, spectral, EPR and in vitro—antibacterial and antioxidant activity. Journal of the Chinese Chemical Society. 2010;**57**(4A):677-682

[75] Adinarayana Reddy S, Janardhan Reddy K, Lakshmi Narayana S, Sarala Y, Varada RA. Synthesis of new reagent 2,6-diacetylpyridine bis-4-phenyl-3- thiosemicarbazone (2,6-DAPBPTSC): Selective, sensitive and extractive spectrophotometric determination of Co(II) in vegetable, soil, pharmaceutical and alloy samples. Journal of the Chinese Chemical Society. 2008;**55**(2):326-334

[76] Matesanz AI, Cuadrado I, Pastor C, Souza P. A novel sulfur-bridged dimeric zinc(II) complex with 2,6-diacetylpyridine bis(thiosemicarbazone). Zeitschrift für Anorganische und Allgemeine Chemie. 2005;**631**(4):780-784

[77] Geng PF, Liu XQ, Zhao TQ, Wang CC, Li ZH, Zhang J, et al. Design, synthesis and in vitro biological evaluation of novel [1,2,3]triazolo[4,5-d] pyrimidine derivatives containing a thiosemicarbazide moiety. European Journal of Medicinal Chemistry. 2018;**146**:147-156

[78] Bernhardt PV, Sharpe PC, Islam M, Lovejoy DB, Kalinowski DS, Richardson DR. Iron chelators of the dipyridylketone thiosemicarbazone class: Precomplexation and transmetalation effects on anticancer activity. Journal of Medicinal Chemistry. 2009;**52**(2):407-415

[79] Sartorelli AC, Booth BA. Inhibition of the growth of sarcoma 180 ascites cells by combinations of inhibitors of nucleic acid biosynthesis and the cupric chelate of Kethoxal bis-(thiosemicarbazone). Cancer Research. 1967;**27**(9):1614-1619

[80] Thiosemicarbazide|CH_5N_3S—PubChem [Internet]. https://pubchem.ncbi.nlm.nih.gov/compound/Thiosemicarbazide. [cited August 23, 2022]

[81] Matesanz AI, Souza P. Unprecedented Pt(II) complex of an asymmetric 2,6-diacetylpyridine bis(4N-substituted thiosemicarbazone) ligand. Inorganic Chemistry Communications. 2013;**27**:5-8

[82] Matesanz AI, Hernández C, Souza P. New bioactive 2,6-diacetylpyridine bis(p-chlorophenylthiosemicarbazone) ligand and its Pd(II) and Pt(II) complexes: Synthesis, characterization, cytotoxic activity and DNA binding ability. Journal of Inorganic Biochemistry. 2014;**138**:16-23

[83] Matesanz AI, Hernández C, Souza P. A new organometallic palladium(II) compound derived of 2,6-diacetylpyridine mono(thiosemicarbazone): Synthesis, spectroscopic properties and crystal structure of two solvatomorphic forms. Journal of Organometallic Chemistry. 2014;**751**:374-378

[84] Matesanz AI, Leitao I, Souza P. Palladium(II) and platinum(II)

bis(thiosemicarbazone) complexes of the 2,6-diacetylpyridine series with high cytotoxic activity in cisplatin resistant A2780cisR tumor cells and reduced toxicity. Journal of Inorganic Biochemistry. 2013;**125**:26-31

[85] Li MX, Chen CL, Zhang D, Niu JY, Ji BS. Mn(II), Co(II) and Zn(II) complexes with heterocyclic substituted thiosemicarbazones: Synthesis, characterization, X-ray crystal structures and antitumor comparison. European Journal of Medicinal Chemistry. 2010;**45**(7):3169-3177

[86] Pedrido R, Bermejo MR, Romero MJ, Vázquez M, González-Noya AM, Maneiro M, et al. Syntheses and X-ray characterization of metal complexes with the pentadentate thiosemicarbazone ligand bis(4-N-methylthiosemicarbazone)-2, 6- diacetylpyridine. The first pentacoordinate lead(II) complex with a pentagonal geometry. Dalton Transactions. 2005;(3):572-579

[87] Li MX, Yang M, Niu JY, Zhi ZL, Qiang XS. A nine-coordinated bismuth(III) complex derived from pentadentate 2,6-diacetylpyridine bis(4N-methylthiosemicarbazone): Crystal structure and both in vitro and in vivo biological evaluation. Inorganic Chemistry. 2012;**51**(Iii):12521-12526

[88] Brađan G, Čobeljić B, Pevec A, Turel I, Milenković M, Radanović D, et al. Synthesis, characterization and antimicrobial activity of pentagonal-bipyramidal isothiocyanato Co(II) and Ni(II) complexes with 2,6-diacetylpyridine bis(trimet hylammoniumacetohydrazone). Journal of Coordination Chemistry. 2016;**69**:801-811

[89] Filipović N, Todorović T, Radanović D, Divjaković V, Marković R, Pajić I, et al. Solid state and solution structures of Cd(II) complexes with two N-heteroaromatic Schiff bases containing ester groups. Polyhedron. 2012;**31**:19-28

[90] Chandra S, Sharma AK. Biological and spectral studies of transition metal complexes with a quinquedentate Schiff base, 2,6-diacetylpyridine bis(thiocarbohydrazone). Journal of Coordination Chemistry. 2009;**62**(22):3688-3700

[91] Li S, Khan MH, Wang X, Cai M, Zhang J, Jiang M, et al. Synthesis of a series of novel In(III) 2,6-diacetylpyridine bis(thiosemicarbazide) complexes: Structure, anticancer function and mechanism. Dalton Transactions. 2020;**49**:17207-17220

[92] Yu P, Deng J, Cai J, Zhang Z, Zhang J, Khan MH, et al. Anticancer and biological properties of a Zn-2,6-diacetylpyridine bis(thiosemicarbazone) complex. Metallomics. 2019;**11**:1372-1386

[93] Leovac VM, Češljević VI. Transition metal complexes with thiosemicarbazide-based ligands, part II, synthesis and characterisation of nickel(II), cobalt(II), manganese(II) and zinc(II) complexes with the pentadentate ligand, 2,6-diacetylpyridine bis(S-methylisothiosemicarbazone). Transition Metal Chemistry. 1987;**12**:504-507

[94] Leovac VM, Bogdanović GA, Česljević VI, Divjaković V. Transition metal complexes with thiosemicarbazide-based ligands. XV. A square-pyramidal NiII complex with an asymmetric coordination of 2,6-diacetylpyridine bis(S-methylisothiosemicarbazone). Acta Crystallographica. Section C. 2000;**56**(8):936-938

[95] Gradinaru JI, Malinovskiĭ ST, Popovici MA, Gdaniec M. Template synthesis and crystal structure of an asymmetric square-planar complex:

Pyridine-2,6-dicarbaldehyde-bis (S-methylisothiosemicarbazonato) nickel(II) iodide. Crystallography Reports. 2005;**50**(2):217-223

[96] Leovac VM, Divjaković V, Češljević VI, Fazlić R. Transition metal complexes with thiosemicarbazide-based ligands. Part 45. Synthesis, crystal and molecular structure of [2,6-diacetylpyridine bis(S-methylisothiosemicarbazonato)] di-azide-iron(III). Journal of the Serbian Chemical Society. 2003;**68**:425-433

[97] Lorenzini C, Pelizzi C, Pelizzi G, Predier G. Investigation into aroylhydrazones as chelating agents. Part 3. Synthesis and spectroscopic characterization of complexes of MnII, CoII, NiII, CuII, and ZnII with 2,6=diacetylpyridine bis(benzoylhydrazone) and X-ray structure of aquachloro [2,6-diacetylpyridine]. Journal of the Chemical Society, Dalton Transactions. 1983:721-727

[98] Ferraz KSO, Silva NF, Jeferson G, de Miranda LF, Romeiro CFD, Souza-fagundes EM, et al. Investigation on the pharmacological profile of 2,6-diacetylpyridine bis(benzoylhydrazone) derivatives and their antimony(III) and bismuth(III) complexes. European Journal of Medicinal Chemistry. 2012;**53**(2):98-106

[99] Santos AF, Ferreira IP, Pinheiro CB, Takahashi JA, Teixeira LR, Beraldo H. Silver(I) complexes with 2,6-diacetylpyridine-bis(benzoylhydrazones): Antifungal activity and interaction with DNA. Polyhedron. 2017;**138**:270-276

[100] Bazhenova TA, Mironov VS, Yakushev IA, Svetogorov RD, Maximova OV, Manakin YV, et al. End-to-end azido-bridged lanthanide chain complexes (Dy, Er, Gd, and Y) with a pentadentate Schiff-Base [N_3O_2] ligand: Synthesis, structure, and magnetism. Inorganic Chemistry. 2020;**59**:563-578

[101] Kalita P, Ahmed N, Bar AK, Dey S, Jana A, Rajaraman G. Pentagonal bipyramidal ln(III) complexes containing an axial phosphine oxide ligand: Field-induced single-ion magnetism behavior of the Dy(III) analogues. Inorganic Chemistry. 2020;**59**:6603-6612

[102] Spillecke L, Koo C, Maximova O, Mironov VS, Kopotkov VA, Korchagin DV, et al. Magnetic behavior of the novel pentagonal-bipyramidal erbium(III) complex (Et3NH) [Er(H2DAPS)Cl2]: High-frequency EPR study and crystal-field analysis. Dalton Transactions. 2021;**50**:18143-18154

[103] Gup R, Gökçe C, Dilek N. Seven-coordinated cobalt(II) complexes with 2,6-diacetylpyridine bis(4-hydroxybenzoylhydrazone): Synthesis, characterisation, DNA binding and cleavage properties. Supramolecular Chemistry. 2015;**27**(10):629-641

[104] Bottari B, Maccari R, Monforte F, Ottaná R, Vigorita MG, Bruno G, et al. Nickel(II) 2,6-Diacetylpyridine bis(isonicotinoylhydrazonate) and bis(benzoylhydrazonate) complexes: Structure and antimycobacterial evaluation. Part XI. Bioorganic & Medicinal Chemistry. 2001;**9**:2203-2211

[105] Mazza P, Orcesi M, Pelizzi C, Pelizzi G, Predieri G, Zani F. Synthesis, structure, antimicrobial, and genotoxic activities of organotin compounds with 2,6-diacetylpyridine nicotinoyl- and Isonicotinoylhydrazones. Journal of Inorganic Biochemistry. 1992;**48**:251-270

[106] Croitor L, Cocu M, Bulhac I, Bourosh PN, Kravtsov VC, Petuhov O, et al. Evolution from discrete mononuclear complexes to trinuclear

linear cluster and 2D coordination polymers of Mn(II) with dihydrazone Schiff bases: Preparation, structure and thermal behavior. Polyhedron. 2021;**206**:115329

[107] Danilescu O, Bourosh PN, Petuhov O, Kulikova OV, Bulhac I, Chumakov YM, et al. Crystal engineering of Schiff Base Zn(II) and Cd(II) Homo- and Zn(II)M(II) (M = Mn or Cd) heterometallic coordination polymers and their ability to accommodate solvent guest molecules. Molecules. 2021;**26**:2317

[108] Main M, v, Fritz JS. Chromatographic determination of metal ions after complexation with bis(quaternary ammonium hydrazones) of 2,6-Diacetylpyridine. Analytical Chemistry. 1989;**61**(11):1272-1275

[109] Vojinović-Ješić LS, Češljević VI, Bogdanović GA, Leovac VM, Szécsényi KM, Divjaković V, et al. Transition metal complexes with Girard reagent-based ligands. Part V. synthesis, characterization and crystal structure of pentagonal-bipyramidal manganese(II) complex with 2,6-diacetylpyridine bis(Girard-T hydrazone). Inorganic Chemistry Communications. 2010;**13**(9):1085-1088

[110] Brađan G, Pevec A, Turel I, Shcherbakov IN, Milenković M, Radanović D, et al. Synthesis, characterization, DFT calculations and antimicrobial activity of pentagonal-bipyramidal Zn(II) and Cd(II) complexes with 2,6-diacetylpyridine-bis(trimethylammoniumacetohydrazone). Journal of Coordination Chemistry. 2016;**69**:2754-2765

[111] Anđelković K, Milenković MR, Pevec A, Turel I, Matić IZ, Vujčić M, et al. Synthesis, characterization and crystal structures of two pentagonal-bipyramidal Fe(III) complexes with dihydrazone of 2,6-diacetylpyridine and Girard's T reagent. Anticancer properties of various metal complexes of the same ligand. Journal of Inorganic Biochemistry. 2017;**174**:137-149

[112] Yusnita J, Puvaneswary S, Mohd. Ali H, Robinson WT, Kwai-Lin T. Synthesis, structural characterization and antibacterial activity of 2,6-diacetylpyridine bis(benzenesulfonohydrazide) Schiff bases and their copper(II) complexes. Polyhedron. 2009;**28**:3050-3054

[113] Yusnita J, Puvaneswary S, Abdulla AM, Robinson WT, Ali HM. Synthesis, characterization and X-ray structures of 2,6-Diacetylpyridinebis(benzenesulfonylhydrazide) and its zinc(II) complex. Journal of Chemical Crystallography. 2009;**39**:615-618

[114] Rodić MV, Radanović MM, Gazdić DV, Leovac VM, Barta Holló B, Raičević V, et al. Reactions of copper(II) bromide with 2,6-diacetylpyridine bis(phenylhydrazone) (L)—Molecular and crystal structures of L and its mixed-valence complex [CuIIL2][CuI 2Br4]. Journal of the Serbian Chemical Society. 2022;**87**:1-15

[115] Anacona JR, Moreno A. Paramagnetic complexes of some d10 ions with a Schiff Base ligand derived from 2,3-diamino-1,4-naphthoquinone and diacetylpyridine: Antibacterial studies. Main Group Metal Chemistry. 1999;**22**(9):573-577

[116] Khalf-Alla PA, Hassan SS, Shoukry MM. Complex formation equilibria, DFT, docking, antioxidant and antimicrobial studies of iron(III) complexes involving Schiff bases derived from glucosamine or ethanolamine. Inorganica Chimica Acta. 2019;**492**:192-197

[117] Paolucci G, Stelluto S, Sitran S, Ajò D, Benetollo F, Polo A, et al. Polynuclear transition metal complexes. The reactivity of the new potentially binucleating ligand 2,6-diacetylpyridine-bis(1′-phthalazinylhydrazone) towards NiII, CuII and ZnII salts, and crystal and molecular structure of {bis[2,6-diacetylpyridine-bis(1′-phthalazinylhydrazonato)nickel(II)]} dimeth. Inorganica Chimica Acta. 1992;**193**:57-75

[118] Barta Holló B, Magyari J, Armaković S, Bogdanović GA, Rodić MV, Armaković SJ, et al. Coordination compounds of a hydrazone derivative with Co(III), Ni(II), Cu(II) and Zn(II): Synthesis, characterization, reactivity assessment and biological evaluation. New Journal of Chemistry. 2016;**40**(7):5885-5895

[119] Holló B, Magyari J, Živković-Radovanović V, Vučković G, Tomić ZD, Szilágyi IM, et al. Synthesis, characterisation and antimicrobial activity of bis(phthalazine-1-hydrazone)-2,6-diacetylpyridine and its complexes with Co III, NiII, CuII and ZnII. Polyhedron. 2014;**80**:142-150

[120] Hajari S, Keypour H, Taher M, Hamed S, Farida M, William R. New 15-membered macrocyclic Schiff base ligand: Synthesis some Cd(II), Mn(II) and Zn(II) complexes, crystal structure, cytotoxicity, antibacterial and antioxidant activity. Journal of Molecular Structure. 2022;**1251**:132049

Chapter 6

Schiff Base as Multifaceted Bioactive Core

Ruchi Shivhare, Kishor Danao, Deweshri Nandurkar, Vijayshri Rokde, Ashwini Ingole, Amol Warokar and Ujwala Mahajan

Abstract

Schiff bases are the condensation products of primary amines and carbonyl compounds, which are becoming more and more significant. Schiff bases are imine or azomethine (–C=N–) functional group containing compounds that are produced through a nucleophile addition process. Excellent chelators called Schiff bases have a place in both qualitative and quantitative analysis of metals in aqueous media. Schiff bases were discovered to be auxiliary scaffolds and adaptable pharmacophore for the creation and production of numerous bioactive leads compounds, and this special quality made them accessible for a wide range of biological applications. Schiff bases exhibit significant biological properties including analgesic, anti-inflammatory, antibacterial, anticonvulsant, anti-tubercular, anticancer, antioxidant, anthelmintic antiglycation, and antidepressant activities. In situ cross-linked hydrogel systems are created using the Schiff bases, which are frequently utilized in coordination, organometallic chemistry, and tissue engineering applications. The role of Schiff bases to the design and creation of new lead with potential biological functions is highlighted in this chapter. Researchers' interest in obtaining the most conclusive and suggestive information on the numerous Schiff bases that have been important for therapeutic purposes over the last few decades and their use in coordination complexes has been maintained by this bioactive core.

Keywords: Schiff bases, azomethine, pharmacophore, coordination complexes, biological

1. Introduction

Schiff bases derivatives are novel approaches to researchers for designing heterocyclic/aryl compounds for emergence of new-fangled nature-friendly technology [1]. Schiff bases have been employed as synthons in the generation of various industrial and biological active compounds such as formazans, 4-thiazolidines, benzoxazines, and so on, namely ring closure, cycloaddition, and replacement reactions [2]. It has played an important role in the development of coordination chemistry and was included as key point in the progression of inorganic biochemistry and optical materials [3].

Basically, Schiff bases are the novel approach compounds that possess imine or azomethine (–C=N–) functional group. Hugo Schiff et al. had first reported that Schiff bases are the condensed product of primary amines and carbonyl compounds [4–7]. In coordination chemistry, Schiff bases are most important class of the widely used organic compounds and have widely application in several fields including inorganic, biological, and analytical chemistry. Schiff bases are important in the therapeutic purposes in medicinal and pharmaceutical field due to its broadly biological activity such as analgesic [8–10], anti-inflammatory [11], anticancer [12, 13], antioxidant [14], anthelmintic [15], antimicrobial [16, 17], anticonvulsant [18], antitubercular [19], and so on. The presence of N atom of azomethine that involved to form of H-bond with active residue of the protein in the cell and influences cellular mechanisms [20, 21]. Moreover, Schiff bases also play an intermediately role in synthesis of organic compounds, pigments, dyes, polymer stabilizers [7], and for corrosion inhibitors [22]. However, studies revealed that metal complexes of Schiff bases depict more biological activity than free organic compounds [23]. Abdel-Rahman et al. [24] have testified numerous transition metal complexes of Schiff bases as ligands that must be different biological activities, likewise antifungal, anticancer, antibacterial, and so on. As example, Fe (II) complexes have been designed, developed, and synthesized using different Schiff bases ligands that derived from 5- bromosalicylaldehyde and variety of alpha-amino acids such as L-arginine, L-histidine, L-phenylalanine, L-aspartic acid, and L-alanine. Although, these complexes are tested for their antimicrobial activity against various bacterial species such as Escherichia coli, Pseudomonas aeruginosa, and Bacillus cereus. Therefore, it is found that Fe (II) complexes unveiled strong antibacterial activities as compared with amino acid Schiff bases ligands. Moreover, Fe (II) complexes were well interacted with calf thymus DNA using UV-visible spectroscopy, and agarose gel electrophoresis measurement at pH = 7.2. As a result, these complexes showed constant binding with different DNA [25]. However, complexes of Schiff base ligand 2-[(2-Hydroxy-3-methoxy-benzylidene)-amino]pyridin-3-ol obtained from 2-amino-3-hydroxy-pyridine and 3- methoxysalicylaldehyde with nanosized Fe(II), Cd(II), and Zn(II) metals have been synthesized by sonication method, and all complexes examined for antimicrobial activity against various bacterial species [26]. Recent study revealed that some metal complexes of Schiff bases have greater cytotoxic activity against colon cancerous cell (HCT-116 cell line). Since the middle of the nineteenth century, metal Schiff base complexes have been recognized. Their usage as Schiff base ligands, which are typically monodentate, bidentate, tridentate, tetradentate, etc., and depend on the presence of donor atoms, has been noted in a significant amount of literature. Because they can create stable compounds with transition metals, they are widely used in coordination chemistry [27].

Researchers have focused their research on the chemistry of metal Schiff base complexes with nitrogen and other donor atoms because of their numerous applications in dyes, polymers, enzyme preparation, as well as used as catalyst in various biological systems. This is due to the stability of Schiff base metal complexes as well as biological activity [28], electrochemistry [29], and potential applications in oxidation catalysis [30]. Due to their selectivity, sensitivity, and synthetic flexibility to the central metal atom and the presence of an azomethine group, which aids in elucidating the biological transformation reaction's mechanism, Schiff bases are widely investigated [31].

Some Cu (II), Co (II), and Ni (II) complexes have been described employing Schiff base ligands generated from 2-amino-3-hydroxypyridine and 3-methoxysalicylaldehyde, and their in vitro antibacterial properties against many bacteria and fungi have been examined. These complexes were further examined for DNA binding, and it was discovered that they could attach to DNA in an intercalative way. However, when compared with the clinically used vinblastine standard, the cytotoxicity of these metal Schiff base complexes on different cell lines such as human colon carcinoma cells (HCT-116 cell line) and breast carcinoma cells (MCF-7 cell line) demonstrated effective cytotoxicity against the growth of carcinoma cells [32]. Che et al. [33] stated that complexes of Pt (II) with Schiff base ligands N,N0-bis(salicylidene)-1,2-ethylenediamine (L1), N,N0- bis(salicylidene)-1,3-propanediamine (L2), and N,N0-bis(salicylidene)- 1,1,2,2-tetramethylethylenediamine (L3) revealed the use of vapor-deposited Pt (II)-salen (11) triplet emitters as for efficient electrophosphorescent dyes in multilayer organic light-emitting diode (OLED) devices with a maximum luminous efficiency of 31 Cd A^{-1}. They discovered that the performance of OLEDs utilizing the Schiff base dopant L3 outperforms that of previously reported Pt (II) emitters.

The majority of Schiff bases are created by condensing salicylaldehyde with both aliphatic and aromatic amines. By using the condensation process of salicylaldehyde with substituted anilines and other aromatic amines, Calvin and Bailes [34] described a number of imines.

These compounds' intriguing electronic characteristics were discovered through spectroscopic research. According to reports, the existence of a lone pair of electrons in these compounds may explain why there is a stronger ligation with metal ions. In the most recent years, several such azomethines and their complexes with various transition metals have been recorded in review papers [35, 36].

Similarly, Gao and Zheung [37] synthesized Schiff base ligands by condensation of 2-hydroxyacetophenone with various chiral diamines such as 1,2-diamino-cyclohexane, 1,2-diphenylethylenediamine, and 2,20-diamino- 1,10-binaphthalene, respectively, to investigate the steric, electronic, and geometric effect of a methyl (-CH3) group on an azomethine carbon in asymmetric catalytic reaction. Because of the impact of p-conjugation in such luminous complexes, More et al. [37] identified Ni (II) and Zn (II) salophen complexes as potential nonlinear optical materials. A considerable number of Schiff base metal complexes have been shown to be relatively good biological compound models [38].

Schiff bases have played an important part in the advancement of contemporary coordination chemistry, as well as in the advancement of inorganic biochemistry, catalysis, and valuable materials due to optical and magnetic characteristics [39]. In recent years, light emission or charge transport capacity technology has piqued the interest of electronic devices such as solar cells and active components for image and data treatment storage [40]. Metal complexes of Schiff bases contain distinct metals (paramagnetic) groups ascribed to magnetism [41].

2. Synthetic approach for Schiff base

Any primary amine can react with an aldehyde or a ketone under appropriate circumstances to produce Schiff bases. Hugo Schiff (1834–1915), a great chemist who was honored by having Schiff base named in his honor, was born in Frankfurt/Main, Germany's thriving Jewish community. The Schiff base generally has the following common structure (**Figure 1**). A Schiff base is a nitrogen analog of an aldehyde or ketone in which the carbonyl group (CO) has been replaced by an imine or azomethine group. It is also known as an imine or an azomethine [42].

The library of organic chemistry has a number of amines and carbonyl compounds that make it possible to create Schiff bases with a variety of structural characteristics.

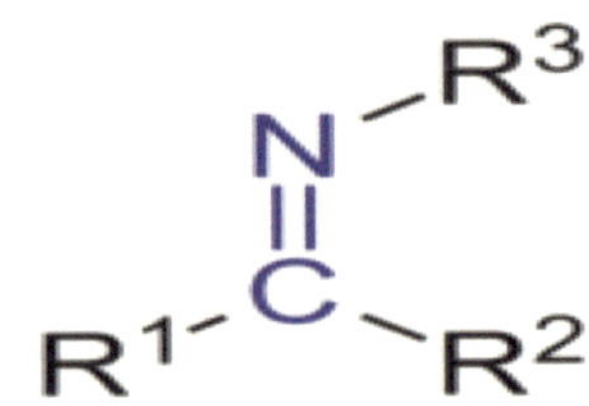

Figure 1.
General structure of Schiff Base (R^1 and R^2 both would be hydrogens. Alkyl or aryl members made up R^3.) [https://en.wikipedia.org/wiki/Schiff_base].

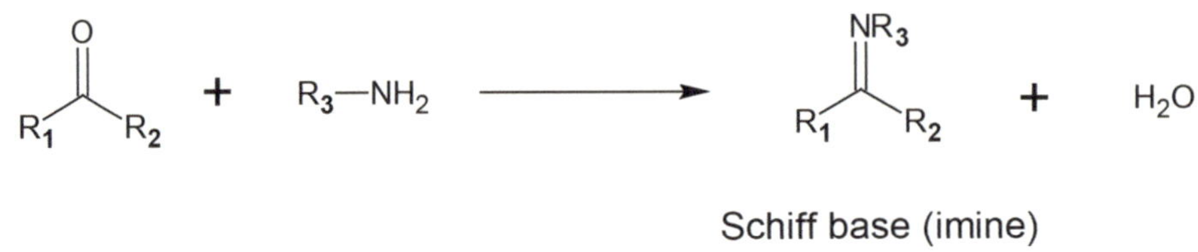

Figure 2.
Schiff base formation reaction scheme [45].

The basic carbonyl group for the production of Schiff bases can be an aldehyde (aromatic or aliphatic) or a ketone [43, 44]. The presence of substituent groups connected to the (>C=N) linkage regulates the stability of the imine group. The general reaction for the synthesis of the Schiff base is shown in (**Figure 2**) where R denotes an alkyl, aryl, cycloalkyl, or heterocyclic group, which might be variably substituted, and R1 may be an alkyl, aryl group, or H atom [46]. Refluxing the mixture under neutral conditions or with acid or basic catalysts usually causes the Schiff base to develop, which is a reversible process. Usually, the product separation or water removal is what brings the formation to completion [47].

Schiff bases are still synthesized by chemists, and today, active and well-designed Schiff base ligands are referred to as "privileged ligands." The bridging Schiff's bases have the following structure, which has a variety of functional groups that can be altered to suit different needs. R" is phenyl or a substituted phenyl, H is an alkyl or aryl group, and X is a phenyl or substituted phenyl group. In fact, Schiff bases have the ability to stable a wide range of metals in a variety of oxidation states, regulating the performance of metals in a wide range of advantageous catalytic transformations. The oxygen atoms in Schiff bases can be changed to sulfur, nitrogen, or selenium atoms; however, NO or N_2O_2-donor groups are the most frequent donor groups [48].

The mechanism of forming a Schiff base (**Figure 3**) is an additional application of the nucleophilic addition to the carbonyl group. The amine is the nucleophile in this situation. The amine interacts with the aldehyde or ketone in the first step of the process to produce carbinolamine, an unstable addition product. By bases or acids catalyzing the process, the carbinolamine loses water. The dehydration of the carbinolamine is acid-catalyzed because it is an alcohol. The reaction is typically catalyzed by acids because the dehydration of the carbinolamine is the rate-determining step in the production of the Schiff base. However, because amines are basic molecules, the acid concentration cannot be too high. The equilibrium is pulled to the left, and carbinolamine production is prevented if the amine is protonated and turns non-nucleophilic. As a result, low acidity is preferred for many Schiff base

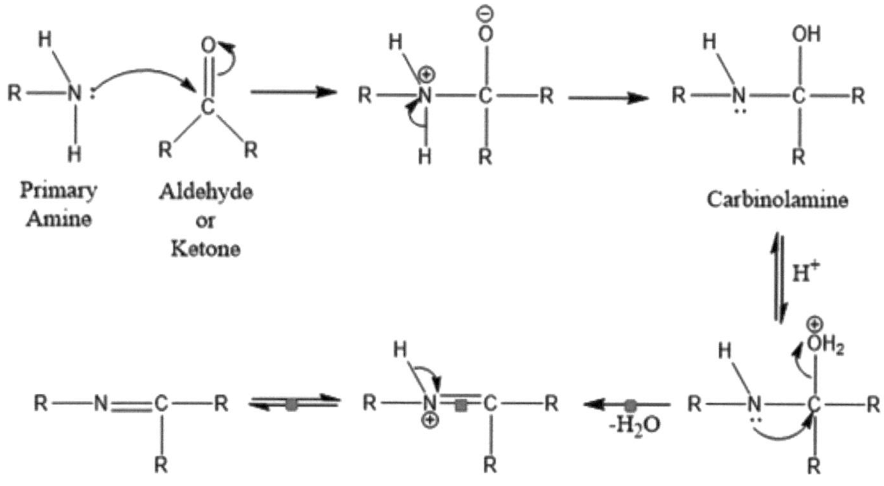

Figure 3.
Mechanism of Schiff base formation [48].

synthess. Base can also catalyze the dehydration of carbinolamines. Despite not being a concerted reaction, this reaction is somewhat comparable to the E2 elimination of alkyl halides. Through an anionic intermediate, it moves forward in two steps. The process of creating a Schiff base actually involves two different types of reactions, addition and elimination [48, 49].

In the meantime, a number of methods and systems for the synthesis of Schiff base have been described, including $NaHCO_3$ [44], $CuSO_4$ [46], P_2O_5/Al_2O_3 [47], $MgClO_4$ [45], $ZnCl_2$ [50], and $MgSO_4$-PPTS [51]. In these systems, metal species act as Lewis acids to activate the carbonyl group and facilitate the removal of water. A few developments have been reported in recent years due to the advancement of experimental procedures, such as solid-state synthesis [52], solvent-free/clay/microwave irradiation [53], water suspension medium [54], reflux/solvent [55], infrared irradiation/no solvent [56], and K-10/microwave. The mentioned methods/systems revealed some drawbacks, including the need for high reaction temperatures, prolonged reaction times, and moisture-sensitive catalysts, huge amounts of aromatic solvents, expensive dehydrating reagents/catalysts, and specialized equipment [57]. $NaHSO_4.SiO_2$/ microwave/solvent-free [58], dirhodium caprolactamate [59], [bmim] BF_4/molecular sieves [60], silica/ultrasound irradiation [61], silica/microwave [19], and silica/solvent-free [62].

Conventional Method for synthesis of Schiff Base: In order to make Schiff bases, aldehyde (0.004 mol) and different aromatic amines (2a–e) (0.004 mol) were condensed in water (10 ml) and agitated at room temperature. TLC has taken note of the reaction's progression. After the reaction was complete, a yellow-color, amorphous product was left behind. This product was filtered, dried, and then recrystallized from methanol [48].

Microwave synthesis for Schiff Base: A mixture of aldehyde (0.004 mol) and substituted aromatic amines (0.004 mol) in water (1 ml) was added and microwaved at 200 W for 30–60 seconds. TLC made note of how the reaction was progressing. Following the completion of the reaction, the reaction mixture contained a solid product, which was filtered and recondensed with methanol [48].

3. Spectroscopic analysis of Schiff base

A Schiff base is a compound with the general structure $R_1R_2C=NR'$. They can be considered a subclass of imines, which is also synonymous with azomethine [63]. In order to investigate hybrid composites, spectroscopic analysis is used. The analysis reveals useful details such as elemental type, chemical composition, optical and electronic properties, and crystallinity. Ultraviolet and visible light (UV-vis) spectroscopy, elemental analysis, differential scanning calorimetric (DSC), hydrogen nuclear magnetic resonance (1H NMR), and Fourier transform infrared (FTIR) studies are used to evaluate these Schiff bases [64].

In the IR spectra, C=N is most commonly reported in the 1690–1640 cm^{-1} region as a strong and a sharp band at somewhat lower frequencies than the bands of C=O groups and close to C=C stretching frequencies. With angle strain, steric repulsion, other complicated local factors, solution concentration and nature of solvent, the stretching frequency of C=N is found to be at 1670 cm^{-1}. The frequency is usually lowered in the absence of one or more groups in conjugation with the C=N [65]. The multinuclear (^1H and ^{13}C) NMR spectral analyses are helpful to characterize and confirm the structure of Schiff bases. The upfield and downfield shifting of the signal is dependent upon the substituents present over the Schiff bases. In the CHN analysis of the Schiff bases, the elemental and sometimes isotopic compositions were found out for the confirmation of the structure of the synthesized derivatives.

It is important to note that the nitrogen atom in the Schiff base has a lone pair, which gives it the characteristics of a Lewis base and allows it to participate in the creation of hydrogen bonds, either intramolecularly or with polar molecules. This characteristic encourages the development of intramolecular hydrogen bonding, particularly when suitable non-polar solvents are present [66].

4. Schiff-base metal complexes

Schiff-base metal complexes have been considered the active topic of research in coordination chemistry during a few decades of extensive research on metal-based pharmaceuticals, owing to their useful applications in numerous disciplines of science. They have potential therapeutic applications as antibiotics, antimicrobials, antitumors, antivirals, anti-inflammatory medicines, analgesics, antifungals, and many others [47]. Schiff bases are versatile pharmacophores that trap in metal ions within their structural units due to the presence of multiple donor atoms [67]. Schiff bases (azomethines) are formed by combining amino and carbonyl groups with multidentate ligands and forming highly significant complexes with metal ions. By using azomethine nitrogen, they can coordinate with metal ions. In organic synthesis, the Schiff base reaction is fundamental for the synthesis of C-N bonds. (**Figure 4**) They exhibit chelation property with O, N, and S donors, and metal complexes have a diverse biological action against many infections and cancers. Schiff base complexes with multidentate ligands are capable of chelating any metal ions. These ligands are effective in the exciting unique therapeutic approach to better understanding diseases and their therapy. The complexes of Schiff base of both transition metal ions (i.e., inner and outer) containing NO or NOS donor atoms were described as playing a significant role in biological activities. Because they are colorful and very stable for biological activities, certain of these metal complexes have attractive physical and chemical properties [69]. Some of the recently synthesized Schiff-base metal complexes are enlisted in **Table 1**.

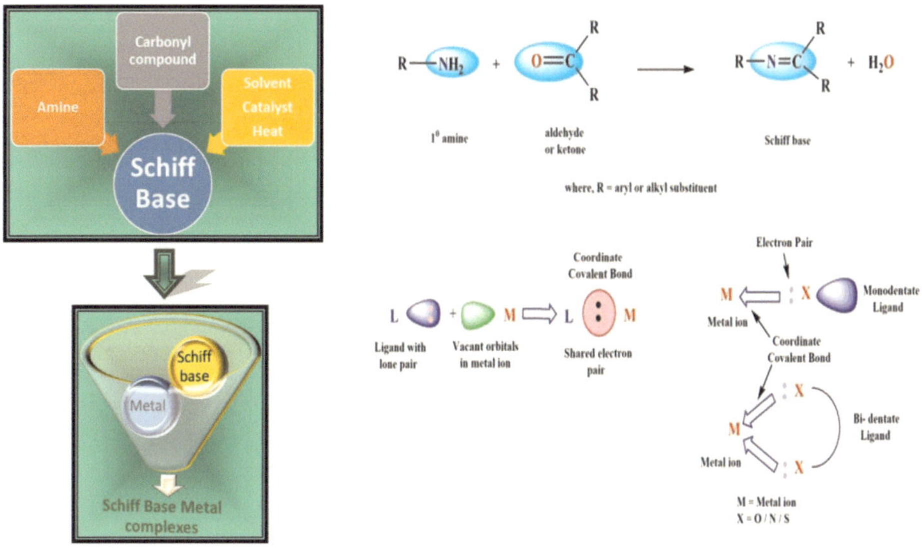

Figure 4.
Formation of Schiff-base metal complexes [68, 69].

5. Schiff base biological activity

Bacteria that display multiple antibiotic resistance [77] are closely linked to the rise in the mortality rate of infectious illnesses [78]. The main reason for this issue is the lack of efficient therapies. There is unquestionably an urgent medical need for the creation of new antibacterial drugs with inventive and more effective modes of action. Hence, Schiff bases called azomethines containing nitrogenous analogue of carbonyl compound changed by imine group are used. For antimicrobial activity, these bases produce excellent activity [79]. Aside from natural and synthetic molecules, biomolecules, sulfonamides, coumarins, aminothiazolyl, bromocoumarins, O-phthaldehyde, or 2-aminophenol, and 1, 2, 4-triazoles are other molecules that can be used as platforms to create Schiff bases for antibacterial activities [80].

As malaria is a disorder associated with severe health problems of public and globally its cases are increasing. Hence, various authors synthesized antimalarial Schiff base derivatives for prominent activity [81]. Also these bases have maximum activity for fungal infection due to its structure. In case of viral disorder, in the market there are numbers of drugs, but they are not fully active due to mutations in viruses. To overcome these problems, Schiff base derivatives are synthesized as an antiviral agent [82]. Different biological activities of Schiff bases are shown in **Table 2**.

6. Schiffs base abiological activity

Schiff base possesses various medicinal activities. In addition to the same, they are now being used in diagnostics also. Some of the abioloical applications of Schiff's base are mentioned in this section.

Sr. no.	Schiff base metal complexes	Work done / Scheme	References
	Influence of co-ligand on the biological properties of Schiff base metal complexes: Synthesis, characterization, cytotoxicity, and antimicrobial studies.	In this work, six new mixed ligand Schiff base metal complexes of $[M_2(L^1)(X)_2Cl_4]$ type, (where, M = Co^{II}, Ni^{II}, Cu^{II}, L^1 = Schiff base ligand derived from 1-pyrenecarboxaldehyde and 1,4-bis[3-aminopropyl]piperazine and X = pyrazine-2-carboxylic acid or 2,2′-biimidazole) were synthesized	[70]
	Synthesis, characterization, design, molecular docking, anti COVID-19 activity, DFT calculations of novel Schiff base with some transition metal complexes	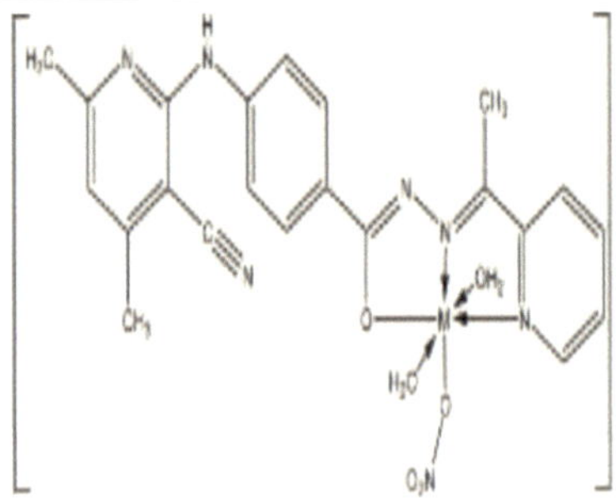 M= Co(II) (n=4); Ni(II) (n=2); Cu(II) (n=0)	[71]
	Synthesis, molecular docking, molecular dynamic, quantum calculation, and antibacterial activity of new Schiff base-metal complexes	Synthesis of Schiff base compounds Ligands (L1 and L2) synthesized from the condensation of Nicotinic acid carbohydrazide with vanillin or 2-chloroquinoline-3-carbaldehyde. The transition metal Co(II) and Cu(II) afforded the Schiff base-metal complexes by metalation.	[72]
	Synthesis, Spectroscopic Characterization and Biological Studies of Mn(II), Cu(II), Ni(II), Co(II) and Zn(II) Complexes with New Schiff Base of 2-((Pyrazine-2-ylimino)methyl)phenol	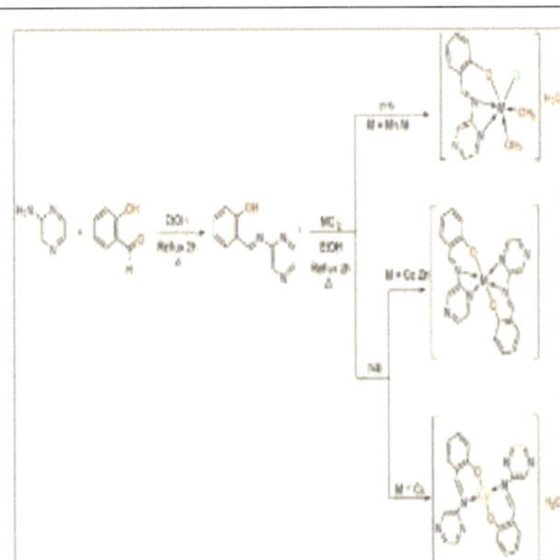	[73]
	Iron (III) and Zinc (II) Metal Alkaloid Complexes: Synthesis, Characterization and Biological Activities	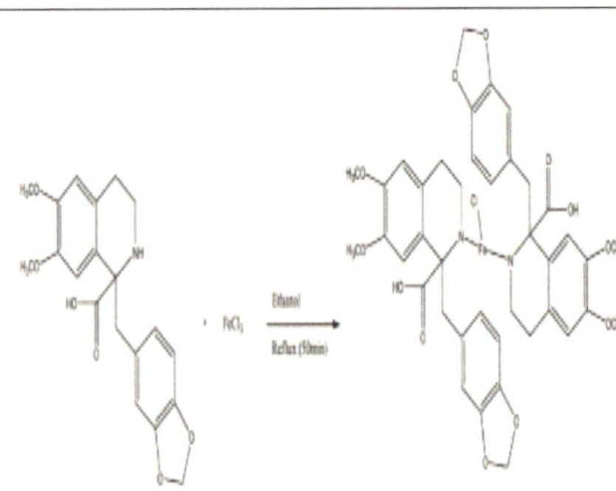	[74]

Sr. no.	Schiff base metal complexes	Work done / Scheme	References
	Novel water soluble Schiff base metal complexes: Synthesis, characterization, antimicrobial-, DNA cleavage, and anticancer activity	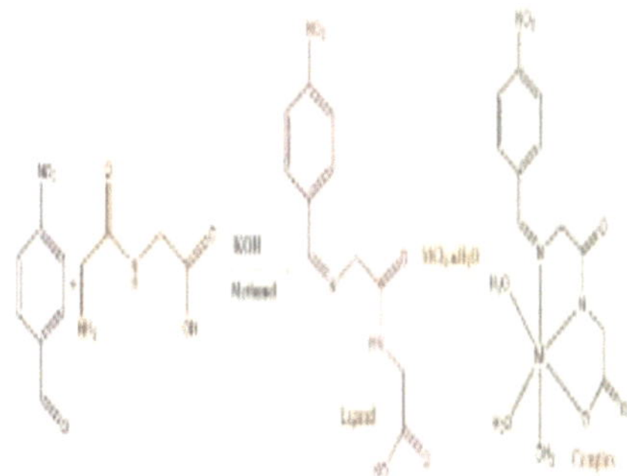	[75]
	Synthesis, characterization and biological studies of Schiff base metal complexes derived from 4-aminoantipyrine, acetamide and p-phenylenediamine	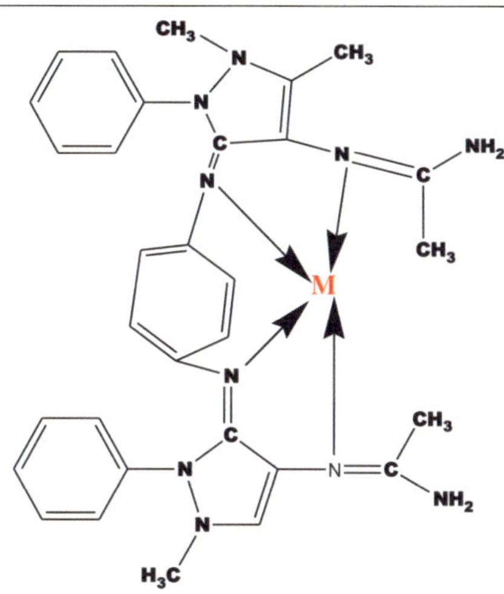	[76]

Table 1.
Recently synthesized Schiff-base metal complex.

6.1 Bioimaging applications

A variety of Schiff base probes with fluorescent sensors are used for bioimaging applications to detect metal ions. Most of the fluorescent probes selectively and sensitively detect only one or two metal ions. Schiff base probes detect analytes in non-aqueous or semi-aqueous media, making them useful for the detection and monitoring of toxic ions in drinking water and industrial waste [100]. A simple and versatile Schiff base chemical sensor was developed to detect four adjacent series 4 metal ions (Co2+, Ni2+, Cu2+, and Zn2+) by colorimetric or fluorometric methods. This chemical sensor has been used to image Zn2+ in HepG2 cells, zebrafish, and tumor-bearing mice, demonstrating potential biological applications [101]. The fluorescence sensor synthesized from the reaction of picolinohydrazide and 4 (diethylamino) salicylaldehyde successfully detected Al3+ and Zn2+ in living cells, suggesting that this simple biosensor has great potential for biological imaging applications [102].

Antimicrobial activity

S.N.	Title	MIC/ IC_{50}	Work done	Structure	Ref
1	Synthesis, Characterization, and Antimalarial Activity of Novel Schiff Base-Phenol and Naphthalene-Amine Ligands	IC_{50} 1.7 micro M	Harpstrite *et al* synthesized Schiff base of phenol and naphthalene by using condensation reactions of 2-Amino-5-diethylamino-pentane and appropriately substituted benzaldehyde or naphthaldehyde in methanol and traerythrocytic culture was used for this acivity by the method of Trager and Jensen for antimicrobial activity CQ-resistant Dd2 strains	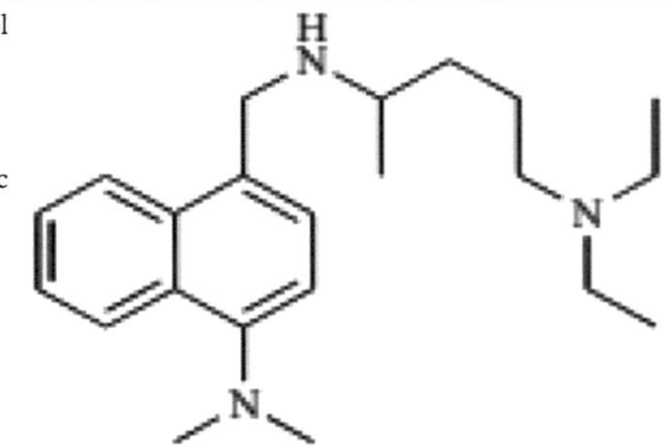	[83]
2	Benzimidazole Schiff base derivatives: synthesis, characterization and antimicrobial activity	MIC 125 μg/mL	Fonkui *et al* synthesized benzimidazole Schiff base derivatives for that Oct-2-ynoic acid (1,3-dihydrobenzoimidazole-2-ylidene)amide was used as starting material which was mixed with different derivatives of aniline in hot ethanol using broth micro dilution technique in which six Gram-positive and eight Gram-negative bacteria were used, authors performed different activity on synthesized derivative as antifungal by using different fungi and different ligands concentrations, anti-malarial and antiparasitic activity. From the synthesized derivatives those containing hydroxyl group, carboxyl group and sulfonyl group showed highest MIC.	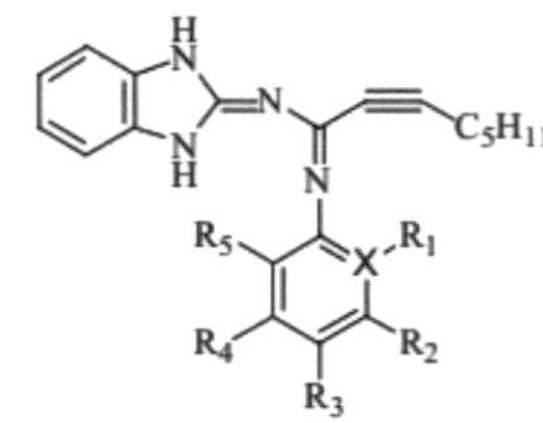	[84]

Antimicrobial activity

S.N.	Title	MIC/ IC_{50}	Work done	Structure	Ref
3	Synthesis, characterization, in vitro antimicrobial, antioxidant and anti-inflammatory activities of diorganotin (IV) complexes derived from salicylaldehyde Schiff bases	25 μg/mL	Devi *et al* synthesized Diorganotin (IV) with Schiff base as complexes were promising areas for antimicrobial activity in coordination chemistry. For MIC determination Serial dilution method was used. Synthesized derivative 20 shows highest MIC.	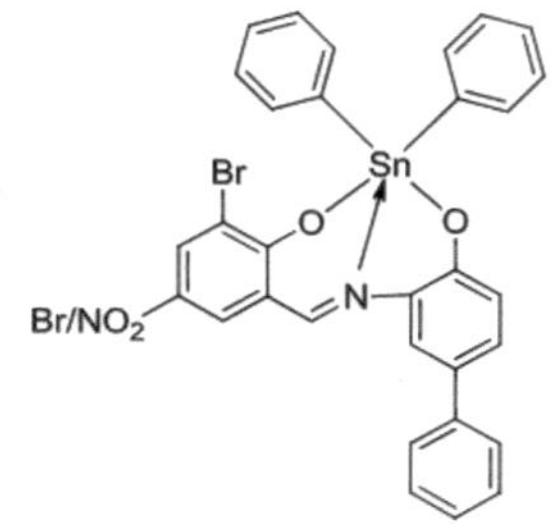	[85]
4	Synthesis, spectral and antimicrobial studies of amino acid derivative Schiff base metal (Co, Mn, Cu, and Cd) complexes	MIC 11.47 μg/mL	Pervaiz *et al* synthesized leucine and salicyldehyde and were combined to create the amino acid derivative Schiff base in a basic media. Antimicrobial and antifungal activity were carried out by using standard Rifampicin and Fluconazole respectively. Mn (L) showed highest MIC as standard showed 21 μg/mL	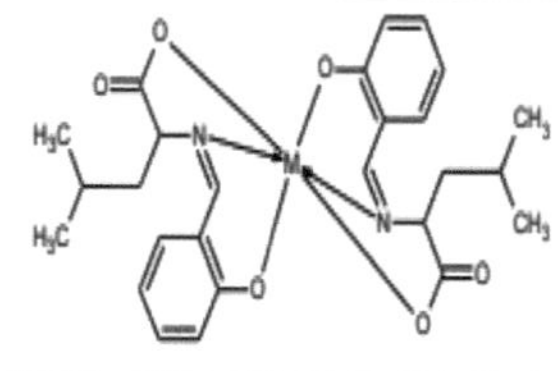	[86]
5	Pyrazole Schiff base hybrids as anti-malarial agents: Synthesis, In vitro screening and computational study	EC50 (6bf) 1.95 μg/ml and (6bd) 1.98 μg/ml	Agrawal *et al* synthesized pyrazole Schiff base for that 2-hydrazinobenzothiazole was used as starting material. Antimalarial activity was done in vitro by using Schizont Maturation Inhibition method against P. falciparum by utilizing RPMI 1640 to grow in human A (+) red blood cells medium and AB (+) serum gentamycin sulfate. Among all derivatives 6bf and 6bd shows highest EC_{50} value having p -CH_3Ph group and meta and para -ClPh group	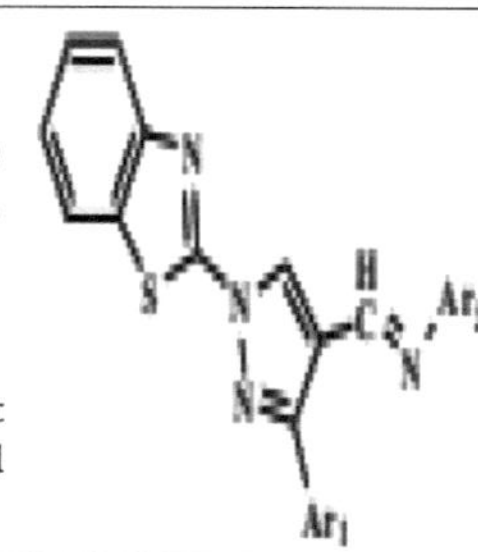	[87]

Antimicrobial activity

S.N.	Title	MIC/ IC_{50}	Work done	Structure	Ref
6	Schiff Bases: A Versatile Pharmacophore		Prakash *et al* synthesized Schiff bases and mannish bases from 5-fluoro isatin for antimicrobial activity by cup plate method from that substituted with electron-donating group of 7-(4-((3-(4-(4Hydroxybenzylideneamino) phenylimino)-5-fluoro-2-oxindolin-1-yl) methyl) piperazin-1-yl)-1-cyclopr opyl-6-fluoro-4-oxo-1,4-dihydroquinoline-3-carboxylic acid (3c) compound shows highest MIC compared to Ciprofloxacin and Ketoconazole. Those derivatives having electron withdrawing effects like nitro and chloro group showed lesser inhibition.	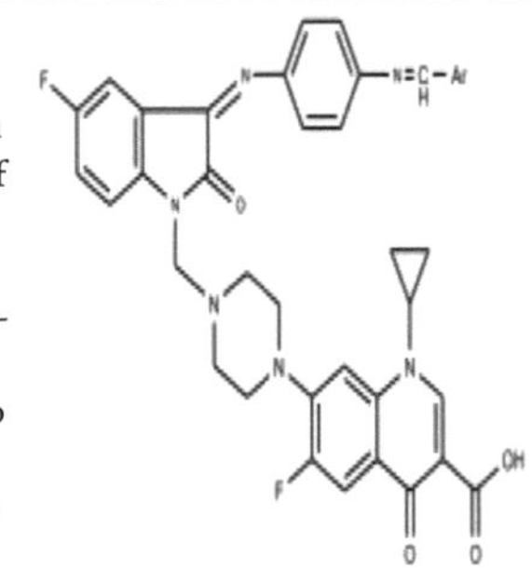	[88]
7	Novel Hydralazine Schiff Base Derivatives and Their Antimicrobial, Antioxidant and Antiplasmodial Properties	MIC 31.25 µg/ml	Sathe *et al* synthesized hydralazine Schiff base derivatives such as 3-[1-(2-(phthalazin-1-yl) hydrazono)ethyl]-1,3-oxa- zinane (PHEO) and 2-[(2-(phthalazin-1 yl) hydrazono) methyl] phenol (PHMP), from that only PHMP showed higher activity	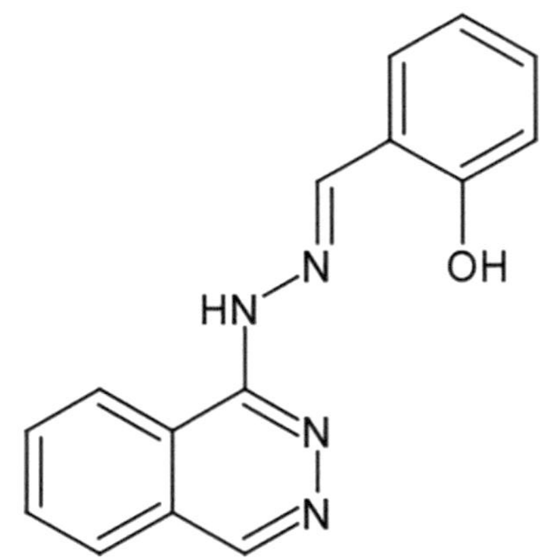	[7]

Antimicrobial activity				
S.N.	**Title**	**MIC/ IC_{50} Work done**	**Structure**	**Ref**
Other activities				
S.N	Title	Work done		
8	Synthesis, characterization, and analgesic activity of novel Schiff base of Isatin derivatives	Chinnasamy *et al* synthesized Isatin Schiff bases by using imesatin and different aromatic aldehydes. Analgesic activity was done by using tail immersion method by using swiss mice having 200 mg/kg body weight and standard drug used was pentazocin. From the synthesized derivatives compound having electron donating groups shows higher analgesic activity such as -(4-hydroxy-3-methoxybenzyllideneamino) phenylimino) indoline-2-one (5i) compound	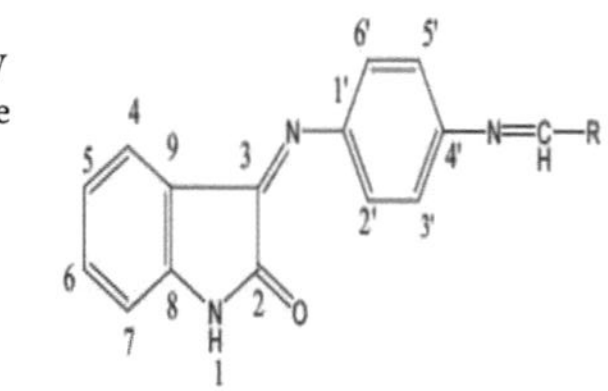	[11]
9	Synthesis, antioxidant and analgesic activities of Schiff bases of 4-amino-1, 2, 4-triazole derivatives containing a pyrazole moiety	Karrouchi *et al* synthesized pyrazole Schiff base and by using Tail flick method analgesic activity was carried out using morphine as standard drug and for its evaluation g acetic acid induced writhing method with albino mice was used. In vitro antioxidant activity was carried out by using DPPH (2, 2-diphenyl-1-picrylhydrazyl) radical. 5 h derivatives having fluoro group showed better analgesic activity.	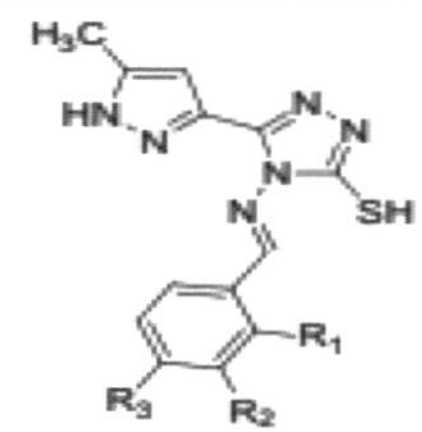	[89]
10	Synthesis, anti-inflammatory, analgesic and kinase (CDK-1, CDK-5 and GSK-3) inhibition activity evaluation of benzimidazole/ benzoxazole derivatives and some Schiff's bases	Sondhi *et al* synthesized Schiff base derivatives of benzimidazole and benzoxazole, anti-inflammatory activity was carried out in the carrageenin-induced paw oedema model in albino rats with 50 kg/mg. Chloro group containing derivatives shows higher activity, derivative naming 4, 4' bis [1-{(2-aminophenylimino) methyl] naphthalen-2-ol showed good anti-inflammatory activity.	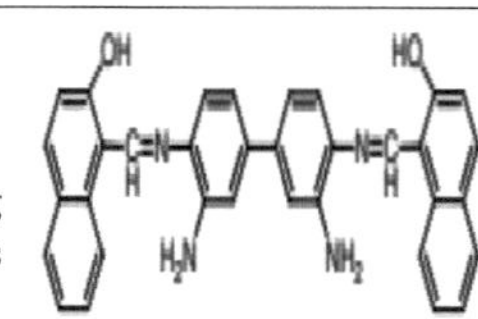	[90]
11	Synthesis and biological evaluation of Schiff bases of 4-aminophenazone as an anti-inflammatory, analgesic and antipyretic agent	Murtaza *et al* synthesized 4-aminophenazone Schiff base, anti-inflammatory activity was done by Carrageenan-induced paw oedema method using albino mice with piroxicam as standard drug. Antipyretic studies was carried through Yeast induced hyper pyrexia (YIHP) in rabbits. Derivative 4-{[(4-bromophenyl)methylidene]amino}-1,5-dimethyl-2-phenyl-	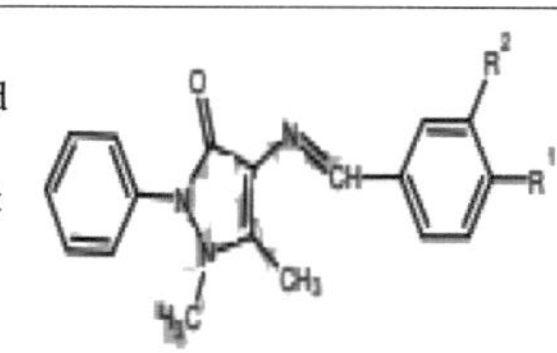	[91]

Antimicrobial activity

S.N.	Title	MIC/ IC_{50} Work done	Structure	Ref
		1,2-dihydro-3H-pyrazol-3-one (4APZAB) showed greater activity than others		
12	Synthesis, characterization, and anticancer activity of Schiff bases	Uddin N synthesized Schiff base by condensation reaction between benzaldehydes and amines in ethanol. Human cervical cancer cells HeLa and human breast adenocarcinoma cells MCF-7 were used in an MTT (dimethyl-2-thiazolyl-2, 5-diphenyl-2H-tetrazolium bromide) based cell viability assay to assess the derivatives potential for cytotoxicity Carboplatin as standard drug. Derivative L5 (5-(diethylamino)-2-((2,6 diethylphenylimino) methyl) phenol showed significant activity	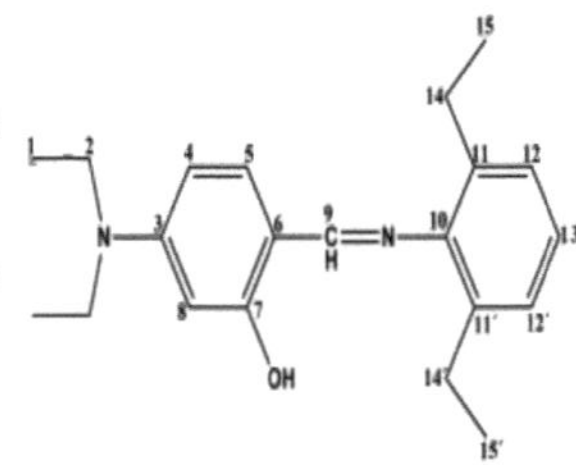	[92]
13	Synthesis, anticancer activity and molecular docking study of Schiff base complexes containing thiazole moiety	Abd-Elzaher *et al* synthesized Schiff base by condensation reaction between salicylaldehydes and thiazole moiety. Different human tumor cell lines, including breast cancer MCF-7, liver cancer HepG2, lung carcinoma A549, and colorectal cancer HCT116, were demonstrated by synthetic Schiff base derivatives in contrast to the activity of doxorubicin as a reference medication.	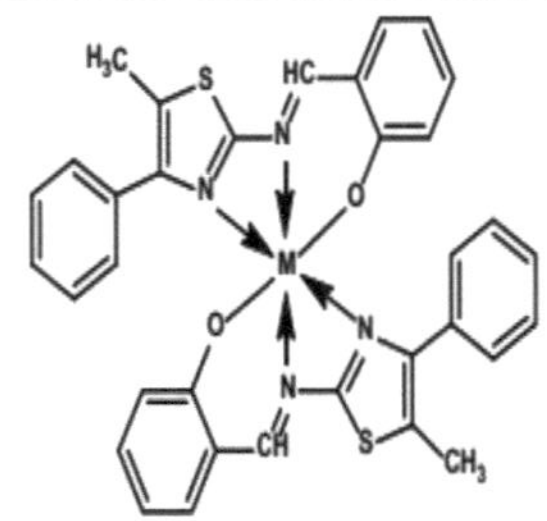	[93]
14	Schiff Bases and their Metal Complexes as Potential Anticancer Candidates: A Review of Recent Works	Tadele *et al* reviewed Schiff base with metal complexes as anticancer agents includes activity as anti-HepG2 and anti-MCF-7 cell activity of quinazolines was extremely high. The strongest activity against various cancer cells was shown by [Ni (HL1)2(OAc) 2], among pyrazole-naphthalene derivatives. Azosal and its tin (IV) complexes shown outstanding efficacy against HCT-116 cell lines and significant activity against U-1242 MG. High activity was shown by 2-thiouracil sulfonamides against MCF7 and CaCo-2 cancer cells. In the presence of visible light, vitamin B6 and its oxovanadium complex shown good action against cervical cancer HeLa, MCF-7, and 3 T3 cell lines. High activity was	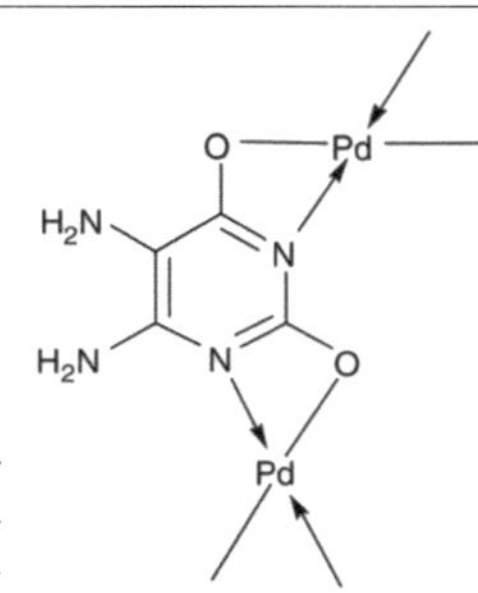	[94]

Antimicrobial activity

S.N.	Title	MIC/ IC_{50} Work done	Structure	Ref
		shown by Indoles against AMJ13. While its binuclear (Y and K) complexes showed high activity against various cancer cells, porphyrine derivatives showed good action. The [Pd(II) and Pt(II)] chitosan complexes exhibited very significant anticancer activity against MCF-7 carcinoma cell.		
15	Schiff Bases: Interesting Scaffolds with Promising Antitumoral Properties	Iacopetta *et al* reviewed antitumor activity of synthesized Schiff base derivatives as mono Schiff derivatives were useful in leukemia cell line. 2, 4 dinitro substituted Schiff base against Human leukemia (HL-60). Also reviewed Potato disc tumor induction assay using Agrobacterium tumefaciens method for this ferrocene containing Schiff base showed highest inhibition of tumor growth. Surfactants used to fight colon cancer (HCT-116), breast cancer (MCF-7), and liver cancer (HepG2) cell lines all include Schiff bases with various lengths of hydrocarbon chains and from that C14 chain containing carbon showed highest inhibition. Anthracene-containing Schiff bases used in human epithelial cancer cell lines. For esophageal cancer cell lines an imine moiety bases were used. Phenylimino-1, 2-diphenylethanol derivatives were used in cell lines of human colon carcinoma and breast carcinoma. Pyranoquinolinone moiety containing Schiff bases used breast cancer cell lines.	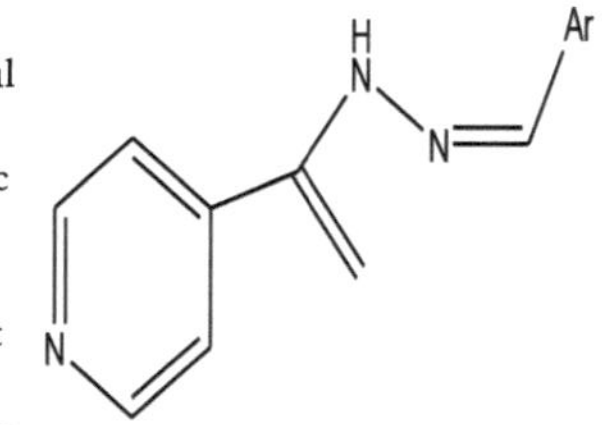	[95]
16	Synthesis and biological evaluation of Schiff's bases and 2-azetidinones of isonocotinyl hydrazone as potential antidepressant and nootropic agents	Thomas *et al* synthesized Schiff bases by cyclocondensation reactions in microwave synthesizer. Synthesized derivatives were tested on swiss albino mice for nootropic activity by elevated plus maze test method. For antidepressant activity Forced swim test and tail suspension test in mice were carried out. Authors interpreted that the studied substances, including 2-azetidinones and Schiff's bases with electron-withdrawing (nitro, halogen, and dimethoxy) moiety on the phenyl ring, were found to have significant *in vivo* nootropic and antidepressant actions in the animal models.	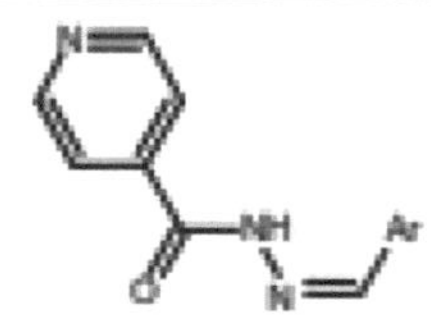	[96]

Antimicrobial activity					
S.N.	Title	MIC/ IC_{50}	Work done	Structure	Ref
17	The Synthesis of Schiff bases and new secondary amine derivatives of p-vanillin and evaluation of their neuroprotective, antidiabetic, antidepressant and antioxidant potentials		*Yuldasheva et al* synthesized Schiff base derivatives of p-vanillin. Neuroprotective by using standard drug neostigmine and galantamine synthesized derivatives of p-vanillin with dopamine and with 4-aminomorpholine showing prominent activity. Acarbose as standard for antidiabetic and trolox for antioxidant activity for them p-vanilline with dopamine showed best among the other derivatives. That means this derivative was most potent inhibitors of glycosidase function.	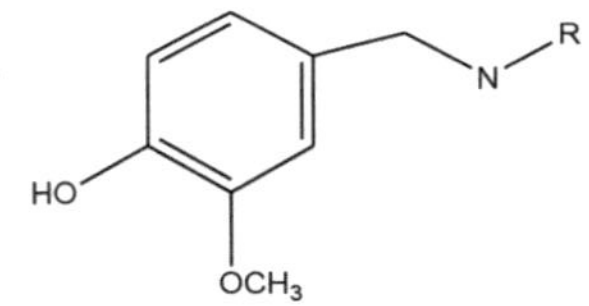	[97]
18	Synthesis and reactions of thiosemicarbazides, triazoles, and Schiff bases as antihypertensive alpha-blocking agents		Authors prepared derivatives by condensing the hydrazide with aldehydes, 2, 3-pyridinedicarboxylic anhydride, or 1, 8-naphthalenedicarboxylic anhydride, Schiff bases were created. For testing white guinea pigs were used. It revealed that several of these chemicals have low toxicity and good hypertension –blocking efficacy.	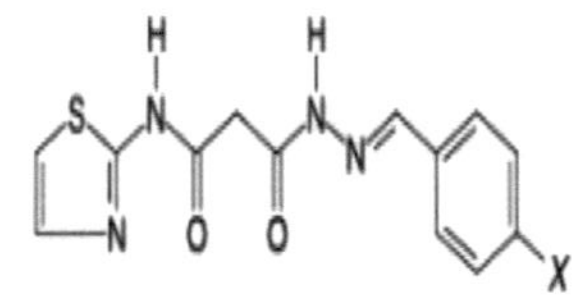	[98]
19	Design, synthesis, spectroscopic characterization and anti-psychotic investigation of some novel Azo dye/Schiff base/Chalcone derivatives.		Gopi et al. synthesized azo, Schiff base and chalcone derivatives. Synthesize derivatives GC8 and GC2 has strong catalepsy induction potency carried out by catalepsy test and its score.	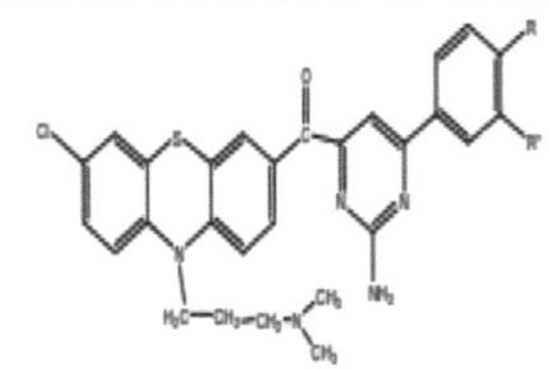	[99]

Table 2.
Different biological activities of Schiff bases.

Compound (E)-1-((L-glutaminoimino) methyl)naphthalene-2-ol (A) showed good solubility and compatibility in the presence or absence of Al3+ and showed some fluorescence in human Hs27 epithelial cells. Bioimaging has been reported [4]. Fluorescent Schiff base organotin dyes (1: Et2N-L-SnPh2, 2: Et2N-L-SnBu2, 3: MeO-L-SnPh2, 4: MeO-L-SnBu2, 5: HO-L-SnPh2, and 6: HO − L-SnBu2, L = 2-hydroxybenzylidene-4-hydroxybenzhydrazine) showed efficient two-photon excitation (1–4). Two of the compounds (5 and 6) were found to be able to selectively accumulate in HeLa cells, allowing their differentiation from normal cells (periodontal ligament cells) [103, 104].

6.2 Phosphorescent OLEDS

Five one-armed Schiff base ligands HL1, HL2, HL3, HL4, and HL5 were obtained from condensation of various group-substituted salicylaldehydes with aniline and 2,4,5-trifluoroaniline gave. Their platinum(II) complexes Pt(L1)2, Pt(L2)2, Pt(L3)2, Pt(L4)2, Pt(L5)2, and PtL5 DMSO obtained by the metalation of ligands with K2PtCl4 were found to be excellent candidates for phosphorescent OLEDs [105]. The luminescent performance of azomethine zinc complexes in organic light-emitting diodes has been investigated, and the results have shown excellent electroluminescence properties as blue fluorescent light sources [106]. Schiff-base zinc metal complexes have been developed to serve as efficient light-emitting materials for optoelectronic applications such as organic light-emitting diodes. These zinc complexes serve as promising emissive layers for optoelectronic applications [107].

6.3 Sensing applications

Fluorescence on/off sensor of a wide range of Schiff bases is being developed for determination of various analytes, toxic ions, and metallic cations and anions in different types of environmental and biological media [108].

Biosensor: Conductive hydrogels based on graphene oxide, dopamine, and polyacrylamide were prepared using the Schiff bases. The high elongation, toughness, and self-adhesion of conducting hydrogels have provided great advantages as biosensors [109].

In forensics, Schiff bases are primarily used in the analysis of illicit drugs. Chemical reactions with Schiff bases reveal illicit drug production and help determine analytes in confiscated samples [110, 111].

6.4 Tissue regeneration

Various substituted Schiff bases have enhanced bulk modulus of the composite hydrogels and slightly increased the in vitro degradation rate. It also promoted cell adhesion and proliferation and maintains the regular cell morphology of bovine articular chondrocytes, increasing potential applications in cartilage tissue engineering [109].

6.5 Bioprint

Water-soluble hydroxybutyl chitosan (HBC) and chondroitin oxysulfate (OCS) have been used to generate bioinks based on the Schiff base reaction, using different sacrificial molds in 3D bioprinting techniques to produce different structures of hydrogels. The controllable shape of HBC/OCS bionic hydrogels can be optimized and customized for specific cartilage engineering applications [109].

6.6 Tissue adhesion

Aldehyde groups in hydrogels based on the Schiff reagent can promote the adhesion of hydrogels to surrounding tissues. An injectable double-cross-linked self-healing hydrogel based on dopamine-grafted oxidized sodium alginate (OSA-DA) and polyacrylamide (PAM) for wound healing has been reported. In terms of hydrogen bonding and Schiff base bonding, the self-healing OSA-DA-PAM hydrogel possesses stable mechanical properties such as high tensile strength and elongation. In addition, numerous catechol groups on OSA-DA chains can endow hydrogels with unique cell affinity and tissue adhesion [10]. In situ forming hydrogel, derived from natural polysaccharides through Schiff base reaction, can be modulated and prepared for soft tissue adhesive, hemostasis, or other biomedical applications in future [110].

6.7 Dyes

New complexes of Zn(II), Pd(II), and Pt(II) with Schiff bases are metal salts of 4-(dibutylamino)-2-hydroxybenzaldehyde and 4,5-diaminophthalonitrile. Sensor applications for imaging surface temperature (planar optodes) and monitoring rapid temperature changes (fiber optic microsensors) have been demonstrated. Pt(II) complexes immobilized in gas permeable matrices also turned out to be promising materials for oxygen measurements. [112].

Schiff bases based on salicylaldehyde units and their use as metal-free organic chromophores can be used to sensitized and co-sensitize dye-sensitized solar cells (DSSCs) [113].

Compounds Et2N-L-SnPh2 and MeO-L-SnPh2 act as an excellent staining for cancer cells (HeLa) using two-photon bioimaging and are expected to have biomedical applications [104].

7. Future prospectus

From the discussion in the section of biological evaluation of Schiff bases, it was clear that the pharmacophore possesses various biological activities, which can be again explored more against various diseases. New groups of organic compounds are still being described, the combinations of which may form a group of extremely desirable compounds with higher potential. The biologically active Schiff's base ligands and metal complexes are playing very crucial role in the drug discovery [114]. Medicinal chemists are now interested in developing novel chemotherapeutic Schiff bases and their metal complexes.

Apart from this, conjugated Schiff Bases have been employed in electronics such the organic field-effect transistor, Perovskite solar cells, and electrochromic devices because they offer some intriguing optoelectronic features. These are also employed in the production of covalent organic framework, which is used in the storage of gas [115]. The measurement of pH values has evolved into one of the most essential necessities with the recent advancements in biological and environmental research. Because of their smooth synthetic roots, easily tuneable structural architecture, non-destructive signals of emission, visually differentiable color generation, and capacity for real sample analysis, organic Schiff base compounds and their derivatives have been observed to play crucial roles in determining the pH values of a particular medium [116]. Because of its bioactive core, Schiff bases have a wide range of applications in the chemical, food, coordination, medical, agricultural, and other industries. This body of literature made it very evident that there will be several opportunities for Schiff bases to become active research molecules in the future.

8. Conclusion

Schiff bases are integral core of the organic chemistry having a verity of biological activities. This chapter focused on the novel leads of numerous Schiff bases having potential medicinal activities with lesser side effects. At last decades, the researchers perceived attention toward bioactive core of Schiff bases, which gained wide medicinal interest. This chapter also explored industrial application of Schiff bases. Advances in this era will require access of the structure activity relationship and mechanism of action of the Schiff bases containing compounds.

At present scenario, Schiff bases are perceived importance in biological activity. It is a versatile organic compound, which is synthesized in reaction between amino compound and aldehyde or ketone well known as imine by condensation process. This imine or azomethine functional groups are versatile pharmacophores for design and development of various bioactive lead compounds. Schiff bases are extensively used for biological activities such antimicrobial, anthelmintic, anti-inflammatory, anticonvulsant, antiTB, antineoplastic, and antidepressant activities. Also, Schiff bases have an intermediately compound in synthesis of various organic compounds. It is also used as catalyst, pigments, and dye synthesis. Furthermore, Schiff bases are used as corrosion inhibitors. The metal complexes of Schiff bases have diversified biological activities. The present chapter summarized the information with respect to diverse biological activities and depicted the recent synthesized variety Schiff bases as potential bioactive core.

Conflict of interest

We confirm that there is no conflict of interest.

Author details

Ruchi Shivhare*, Kishor Danao, Deweshri Nandurkar, Vijayshri Rokde, Ashwini Ingole, Amol Warokar and Ujwala Mahajan
Dadasaheb Balpande College of Pharmacy, Nagpur, Maharashtra, India

*Address all correspondence to: shivharer4@gmail.com

References

[1] Bhattacharya A, Purohit VC, Rinaldi F. Environmentally friendly solvent-free processes: Novel dual catalyst system in Henry reaction. Organic Process Research and Development. 2003;7(3): 254-258

[2] Jarrahpour A, Khalili D, De Clercq E, Salmi C, Brunel JM. Synthesis, antibacterial, antifungal and antiviral activity evaluation of some new bis-Schiff bases of isatin and their derivatives. Molecules. 2007;**12**(8): 1720-1730

[3] Tisato F, Refosco F, Bandoli G. Structural survey of technetium complexes. Coordination Chemistry Reviews. 1994;**135–136**:325-397

[4] Cimerman Z, Miljanic S, Galic N. Schiff bases derived from aminopyridines as spectrofluorimetric analytical reagents. Croatica Chemica Acta. 2000;**73**(1):81-95

[5] Schiff H. Mittheilungen aus dem Universit¨atslaboratorium in Pisa: eine neue reihe organischer Basen. Justus Liebigs Annalen der Chemie. 1864; **131**(1):118-119

[6] Dhar DN, Taploo CL. Schiff bases and their applications. Journal of Scientific and Industrial Research. 1982;**41**(8): 501-506

[7] Sathe BS, Jaychandran E, Jagtap VA, Sreenivasa GM. Synthesis characterization and anti-inflammatory evaluation of new fluorobenzothiazole schiff's bases. International Journal of Pharmaceutical Research and Development. 2011;**3**(3):164-169

[8] Sondhi SM, Singh N, Kumar A, Lozach O, Meijer L. Synthesis, anti-inflammatory, analgesic and kinase (CDK-1, CDK-5 and GSK-3) inhibition activity evaluation of benzimidazole/ benzoxazole derivatives and some Schiff's bases. Bioorganic and Medicinal Chemistry. 2006;**14**(11):3758-3765

[9] Pandey A, Dewangan D, Verma S, Mishra A, Dubey RD. Synthesis of schiff bases of 2-amino-5-aryl-1,3,4-thiadiazole and its analgesic, anti-inflammatory, anti-bacterial and antitubercular activity. International Journal of ChemTech Research. 2011;**3**(1):178-184

[10] Chandramouli C, Shivanand MR, Nayanbhai TB, Bheemachari B, Udupi RH. Synthesis and biological screening of certain new triazole schiff bases and their derivatives bearing substituted benzothiazole moiety. Journal of Chemical and Pharmaceutical Research. 2012;**4**(2):1151-1159

[11] Chinnasamy RP, Sundararajan R, Govindaraj S. Synthesis, characterization, and analgesic activity of novel Schiff base of isatin derivatives. Journal of Advanced Pharmaceutical Technology and Research. 2010;**1**(3): 342-347

[12] Venkatesh P. Synthesis, characterization and antimicrobial activity of various schiff bases complexes of Zn(II) and Cu(II) ions. Asian Journal of Pharmaceutical and Health Sciences. 2011;**1**(1):8-11

[13] Chaubey AK, Pandeya SN. Synthesis & anticonvulsant activity (chemo shock) of Schiff and Mannich bases of Isatin derivatives with 2-amino pyridine (mechanism of action). International Journal of PharmTech Research. 2012; **4**(4):590-598

[14] Aboul-Fadl T, Mohammed FA, Hassan EA. Synthesis, antitubercular

activity and pharmacokinetic studies of some Schiff bases derived from 1- alkylisatin and isonicotinic acid hydrazide (INH). Archives of Pharmacal Research. 2003;**26**(10):778-784

[15] Miri R, Razzaghi-asl N, Mohammadi MK. QM study and conformational analysis of an isatin Schiff base as a potential cytotoxic agent. Journal of Molecular Modeling. 2013;**19**(2): 727-735

[16] Mounika K, Anupama B, Pragathi J, Gyanakumari C. Synthesis, characterization and biological activity of a Schiff base derived from 3-ethoxy salicylaldehyde and 2-amino benzoic acid and its transition metal complexes. Journal of Scientific Research. 2010;**2**(3): 513-524

[17] Ali SMM, Azad MAK, Jesmin M, et al. In vivo anticancer activity of vanillin semicarbazone. Asian Pacific Journal of Tropical Biomedicine. 2012;**2**(6):438-442

[18] Wei D, Li N, Lu G, Yao K. Synthesis, catalytic and biological activity of novel dinuclear copper complexwith Schiff base. Science in China B. 2006;**49**(3): 225-229

[19] Avaji PG, Vinod Kumar CH, Patil SA, Shivananda KN, Nagaraju C. Synthesis, spectral characterization, in-vitro microbiological evaluation and cytotoxic activities of novel macrocyclic bis hydrazone. European Journal of Medicinal Chemistry. 2009;**44**(9): 3552-3559

[20] Venugopala KN, Jayashree BS. Synthesis of carboxamides of 2-amino-4-(6-bromo-3-coumarinyl) thiazole as analgesic and antiinflammatory agents. Indian Journal of Heterocyclic Chemistry. 2003;**12**(4):307-310

[21] Vashi K, Naik HB. Synthesis of novel Schiff base and azetidinone derivatives and their antibacterial activity. European Journal of Chemistry. 2004;**1**:272-276

[22] Li S, Chen S, Lei S, Ma H, Yu R, Liu D. Investigation on some Schiff bases as HCl corrosion inhibitors for copper. Corrosion Science. 1999;**41**(7):1273-1287

[23] Chohan ZH, Praveen M, Ghaffar A. Structural and biological behaviour of Co (II), Cu(II) and Ni(II) metal complexes of some amino acid derived Schiff-bases. Metal-Based Drugs. 1997;**4**(5):267-272

[24] Abdel-Rahman LH, Abu-Dief AM, El-Khatib RM, Abdel-Fatah SM. Some new nano-sized Fe (II), Cd (II) and Zn (II) Schiff base complexes as precursor for metal oxides: Sonochemical synthesis, characterization, DNA interaction, in vitro antimicrobial and anticancer activities. Bioorganic Chemistry. 1 Dec 2016;**69**:140-152

[25] Malladi S, Isloor AM, Isloor S, Akhila DS. Synthesis, characterization and antibacterial activity of some newpyrazole based Schiff bases. Arabian Journal of Chemistry. 2013;**6**(3):335-340

[26] Essa HH, Kandil F, Falah A. Synthesis and identification of schiff bases and biological activity new study. Iraqi Journal of Science. 2012;**53**(2): 230-240

[27] Shelke VA, Jadhav SM, Patharkar VR, Shankarwar SG, Munde AS, Chondhekar TK. Synthesis, spectroscopic characterization and thermal studies of some rare earth metal complexes of unsymmetrical tetradentate Schiff base ligand. Arabian Journal of Chemistry. 2012;**5**:501-507

[28] Sheikh RA, Wani MY, Shreaz S, Hashmi AA. Synthesis, characterization and biological screening of some Schiff

base macrocyclic ligand based transition metal complexes as antifungal agents. Arabian Journal of Chemistry. 1 Sep 2016;**9**:S743-S751

[29] Sashidhara KV, Kumar A, Bhatia G, Khan MM, Khanna AK, Saxena JK. Antidyslipidemic and antioxidative activities of 8-hydroxyquinoline derived novel keto-enamine Schiffs bases. European Journal of Medicinal Chemistry. 2009;**44**(4):1813-1818

[30] Sashidhara KV, Rosaiah JN, Bhatia G, Saxena JK. Novel keto-enamine Schiffs bases from7-hydroxy-4-methyl-2-oxo-2H-benzo[h] chromene-8,10-dicarbaldehyde as potential antidyslipidemic and antioxidant agents. European Journal of Medicinal Chemistry. 2008;**43**(11):2592-2596

[31] Revanasiddappa HD, Prasad KS, Kumar LS, Jayalakshmi B. Synthesis and biological activity of new Schiff bases containing 4(3H)-quinazolinone ring system. International Journal of ChemTech Research. 2010;**2**(2): 1344-1349

[32] Kundu A, Shakil NA, Saxena DB, Pankaj P, Kumar J, Walia S. Microwave assisted solvent-free synthesis and biological activities of novel imines (Schiff bases). Journal of Environmental Science and Health B. 2009;**44**(5): 428-434

[33] Che CM, Chan SC, Xiang HF, Chan MC, Liu Y, Wang Y. Tetradentate Schiff base platinum (II) complexes as new class of phosphorescent materials for high-efficiency and white-light electroluminescent devices. Chemical Communications. 2004;(13):1484-1485

[34] Aboul-Fadl T, Bin-Jubair FAS, Aboul-Wafa O. Schiff bases of indoline-2,3-dione (isatin) derivatives and nalidixic acid carbohydrazide, synthesis, antitubercular activity and pharmacophoric model building. European Journal of Medicinal Chemistry. 2010;**45**(10):4578-4586

[35] Ferreira MDL, Vasconcelos TRA, de Carvalho EM, et al. Synthesis and antitubercular activity of novel Schiff bases derived from d-mannitol. Carbohydrate Research. 2009;**344**(15): 2042-2047

[36] Thomas AB, Nanda RK, Kothapalli LP, Hamane SC. Synthesis and biological evaluation of Schiff's bases and 2-azetidinones of isonocotinyl hydrazone as potential antidepressant and nootropic agents. Arabian Journal of Chemistry. 1 Sep 2016;**9**:S79-S90

[37] Gao WT, Zheng Z. Synthetic studies on optically active Schiff-base ligands derived from condensation of 2-hydroxyacetophenone and chiral diamines. Molecules. 31 Jul 2002;**7**(7): 511-516

[38] More MS, Devkule SS, Chavan SS. Ni (II) and Zn (II) complexes containing alkynyl functionalized salicylaldimine ligand and heterocyclic coligand: Synthesis, characterization and luminescence properties. Journal of Fluorescence. May 2017;**27**(3):841-851

[39] Aly MM, Mohamed YA, El-Bayouki KAM, Basyouni WM, Abbas SY. Synthesis of some new 4(3H)-quinazolinone-2-carboxaldehyde thiosemicarbazones and their metal complexes and a study on their anticonvulsant, analgesic, cytotoxic and antimicrobial activities e Part-1. European Journal of Medicinal Chemistry. 2010;**45**(8):3365-3373

[40] Nithinchandra B, Kalluraya SA, Shabaraya AR. Regioselective reaction: Synthesis, characterization and pharmacological activity of some new

Mannich and Schiff bases containing sydnone. European Journal of Medicinal Chemistry. 2012;**54**:597-604

[41] Alam MS, Choi J, Lee D. Synthesis of novel Schiff base analogues of 4-amino-1, 5-dimethyl-2- phenylpyrazol-3- one and their evaluation for antioxidant and anti-inflammatory activity. Bioorganic & Medicinal Chemistry. 2012;**20**: 4103-4108

[42] Da Silva CM, da Silva DL, Modolo LV, Alves RB, de Resende MA, Martins CV, et al. Schiff bases: A short review of their antimicrobial activities. Journal of Advanced Research. 2011; **2**(1):1, 1-8

[43] Zheng Y, Ma K, Li H, Li J, He J, Sun X, et al. One pot synthesis of imines from aromatic nitro compounds with a novel Ni/SiO2 magnetic catalyst. Catalysis Letters. 2009 Mar;**128**(3):465-474

[44] Samec JS, Bäckvall JE. Ruthenium-catalyzed transfer hydrogenation of imines by Propan-2-ol in benzene. Chemistry–a. European Journal. 2002; **8**(13):2955-2961

[45] Kumar KS, Bayeh Y, Gebretsadik T, Elemo F, Gebrezgiabher M, Thomas M, et al. Spin-crossover in iron (ii)-Schiff base complexes. Dalton Transactions. 2019;**48**(41):15321-15337

[46] Jennings WB, Lovely CJ. An efficient method for the preparation of N-phosphinoyl and N-sulphonyl imines directly from aromatic aldehydes. Tetrahedron Letters. 1988;**29**(30): 3725-3728

[47] Chaudhary NK, Guragain B, Chaudhary SK, Mishra P. Schiff base metal complex as a potential therapeutic drug in medical science: A critical review. Bibechana. 2021;**18**(1):214-230

[48] Anamika S, Chaturvedi P. Review on Schiff bases. World Journal of Pharmaceutical Sciences. 2021;**27**: 133-140

[49] Naeimi H, Salimi F, Rabiei K. Mild and convenient one pot synthesis of Schiff bases in the presence of P2O5/ Al2O3 as new catalyst under solvent-free conditions. Journal of Molecular Catalysis A: Chemical. 2006;**260**(1–2): 100-104

[50] Billman JH, Tai KM. Reduction of schiff bases. II. Benzhydrylamines and structurally related compounds1a, b. The Journal of Organic Chemistry. 1958; **23**(4):535-539

[51] Chakraborti AK, Bhagat S, Rudrawar S. Magnesium perchlorate as an efficient catalyst for the synthesis of imines and phenylhydrazones. Tetrahedron Letters. 2004;**45**(41):7641

[52] Schmeyers J, Toda F, Boy J, Kaupp G. Quantitative solid–solid synthesis of azomethines. Journal of the Chemical Society, Perkin Transactions. 1998;**2**(4): 989-994

[53] Varma RS, Dahiya R, Kumar S. Clay catalyzed synthesis of imines and enamines under solvent-free conditions using microwave irradiation. Tetrahedron Letters. 1997;**38**(12): 2039-2042

[54] Tanaka K, Shiraishi R. Clean and efficient condensation reactions of aldehydes and amines in a water suspension medium. Green Chemistry. 2000;**2**(6):272-273

[55] Casey CP, Johnson JB. Isomerization and deuterium scrambling evidence for a change in the rate-limiting step during imine hydrogenation by Shvo's hydroxycyclopentadienyl ruthenium hydride. Journal of the American

Chemical Society. 2005;**127**(6): 1883-1894

[56] Vázquez MÁ, Landa M, Reyes L, Miranda R, Tamariz J, Delgado F. Infrared irradiation: Effective promoter in the formation of n-benzylideneanilines in the absence of solvent. Synthetic communications. 2004;**34**(15):2705-2718

[57] Guzen KP, Guarezemini AS, Orfao AT, Cella R, Pereira CM, Stefani HA. Eco-friendly synthesis of imines by ultrasound irradiation. Tetrahedron Letters. 2007;**48**(10):1845-1848

[58] Gopalakrishnan M, Sureshkumar P, Kanagarajan V, Thanusu J, Govindaraju R. Silica gel supported sodium hydrogen sulfate as an efficient and reusable heterogeneous catalyst for the synthesis of imines in solvent-free conditions under microwave irradiation. Journal of Chemical Research. 2005;**2005**(5): 299-303

[59] Choi H, Doyle MP. Oxidation of secondary amines catalyzed by dirhodium caprolactamate. Chemical Communications. 2007;**7**:745-747

[60] Andrade CK, Takada SC, Alves LM, Rodrigues JP, Suarez PA, Brandao RF, et al. Molecular sieves in ionic liquids as an efficient and recyclable medium for the synthesis of imines. Synlett. 2004; **2004**(12):2135-2138

[61] Venugopala KN, Jayashree BS. Microwave-induced synthesis of Schiff bases of aminothiazolyl bromocoumarins as antibacterials. Indian Journal of Pharmaceutical Sciences. 2008;**70**(1):88

[62] Kagatikar S, Sunil D. Schiff bases and their complexes in organic light emitting diode application. Journal of Electronic Materials. 2021;**50**(12): 6708-6723

[63] Ceramella J, Iacopetta D, Catalano A, Cirillo F, Lappano R, Sinicropi MS. A review on the antimicrobial activity of Schiff bases: Data collection and recent studies. Antibiotics. 2022;**11**(2):191

[64] Jarząbek B, Kaczmarczyk B, Sęk D. Characteristic and spectroscopic properties of the Schiff-base model compounds. Spectrochimica Acta. Part A, Molecular and Biomolecular Spectroscopy. 2009;**74**(4):949-954

[65] Rodríguez MR, Del Pla J, Piro OE, Echeverría GA, Espino G, Pis-Diez R, et al. Structure, tautomerism, spectroscopic and DFT study of o-vanillin derived Schiff bases containing thiophene ring. Journal of Molecular Structure. 2018; **1165**:381-390

[66] Raczuk E, Dmochowska B, Samaszko-Fiertek J, Madaj J. Different Schiff bases—Structure, importance and classification. Molecules. 2022;**27**:787

[67] Kajal A, Bala S, Kamboj S, Sharma N, Saini V. Schiff bases: A versatile pharmacophore. Journal of Catalysts. 2013;**2013**:27

[68] Jain A, De S, Barman P. Microwave-assisted synthesis and notable applications of Schiff-base and metal complexes: A comparative study. Research on Chemical Intermediates. 2022;**48**(5):2199-2251

[69] Ghanghas P, Choudhary A, Kumar D, Poonia K. Coordination metal complexes with Schiff bases: Useful pharmacophores with comprehensive biological applications. Inorganic Chemistry Communications. 2021;**130**: 108710

[70] Vijayan T, Kim J, Azam M, Al-Resayes SI, Stalin A, Kannan BS, et al. Influence of co-ligand on the biological properties of Schiff base metal

complexes: Synthesis, characterization, cytotoxicity, and antimicrobial studies. Applied Organometallic Chemistry. 2022;**36**(3):e6542

[71] El-Gammal OA, El-Bindary AA, Mohamed FS, Rezk GN, El-Bindary MA. Synthesis, characterization, design, molecular docking, anti COVID-19 activity, DFT calculations of novel Schiff base with some transition metal complexes. Journal of Molecular Liquids. 2022;**346**:117850

[72] Hashem HE, Nath A, Kumer A. Synthesis, molecular docking, molecular dynamic, quantum calculation, and antibacterial activity of new Schiff base-metal complexes. Journal of Molecular Structure. 2022;**1250**:131915

[73] Howsaui HB, Sharfalddin AA, Abdellattif MH, Basaleh AS, Hussien MA. Synthesis, spectroscopic characterization and biological studies of Mn (II), Cu (II), Ni (II), Co (II) and Zn (II) complexes with new Schiff Base of 2-((Pyrazine-2-ylimino) methyl) phenol. Applied Sciences. 2021;**11**(19):9067

[74] Naureen B, Miana GA, Shahid K, Asghar M, Tanveer S, Sarwar A. Iron (III) and zinc (II) monodentate Schiff base metal complexes: Synthesis, characterisation and biological activities. Journal of Molecular Structure. 2021; **1231**:129946

[75] Shiju C, Arish D, Kumaresan S. Novel water soluble Schiff base metal complexes: Synthesis, characterization, antimicrobial-, DNA cleavage, and anticancer activity. Journal of Molecular Structure. 2020;**1221**:128770

[76] Dhanaraj CJ, Raj SS. Synthesis, characterization and biological studies of Schiff base metal complexes derived from 4-aminoantipyrine, acetamide and p-phenylenediamine. Inorganic Chemistry Communications. 2020;**119**: 108087

[77] Frieri M et al. Antibiotic resistance. Journal of Infectious Public Health. 2016: S18

[78] Kimberly W, John D. Global trends in infectious diseases of swine. Proceedings of the National Academy of Sciences. 2017;**2017**:201

[79] Cinarli A, Gürbüz D, Tavman A, Birteksöz AS. Synthesis, spectral characterizations and antimicrobial activity of some Schiff bases of 4-chloro-2-aminophenol. Bulletin of the Chemical Society of Ethiopia. 2011;**25**(3):407-417

[80] Karrouchi K, Chemlal L, Taoufik J, Cherrah Y, Radi S, El Abbes FM, et al. Synthesis, antioxidant and analgesic activities of Schiff bases of 4-amino-1,2,4-triazole derivatives containing a pyrazole moiety. Annales Pharmaceutiques Françaises. 2016;**74**(6):431-438

[81] Krungkrai J, Krungkrai SR, Supuran CT. Malarial parasite carbonic anhydrase and its inhibitors. Current Topics in Medicinal Chemistry. 2007;**7**(9):909-917

[82] Maryam M, Tan SL, Crouse KA, Mohamed Tahir MI, Chee HY. Synthesis, characterization and evaluation of antidengue activity of enantiomeric Schiff bases derived from S-substituted dithiocarbazate. Turkish Journal of Chemistry. 2020;**44**(5):1395-1409

[83] Harpstrite SE, Collins SD, Oksman A, Goldberg DE, Sharma V. Synthesis, characterization, and antimalarial activity of novel Schiff Base-phenol and naphthalene-amine ligands. Medicinal Chemistry. 2008;**4**:392-395

[84] Fonkui TY, Ikhile MI, Njobeh PB, Ndinteh DT. Benzimidazole Schif base

derivatives: Synthesis, characterization and antimicrobial activity. BMC Chemistry. 2019;**13**:127

[85] Devi J, Yadav J, Singh N. Synthesis, characterisation, in vitro antimicrobial, antioxidant and anti-inflammatory activities of diorganotin (IV) complexes derived from salicylaldehyde Schiff bases. Research on Chemical Intermediates. Jul 2019;**45**(7):3943-3968

[86] Pervaiz M, Ahmad I, Yousaf M, Kirn S, Munawar A, Saeed Z, et al. Synthesis, spectral and antimicrobial studies of amino acid derivative Schiff base metal (Co, Mn, Cu, and Cd) complexes. Spectrochimica Acta. Part A, Molecular and Biomolecular Spectroscopy. 2019; **(5) 206**:642-649

[87] Aggrawal S, Paliwal D, Kaushik D, Gupta GK, Kumar A. Pyrazole schiff base hybrids as anti-malarial agents: Synthesis, In vitro screening and computational study. Combinatorial Chemistry & High Throughput Screening. 2018;**21**:194-203

[88] Prakash CR, Raja S. Synthesis, characterization and in vitro antimicrobial activity of some novel 5-substituted Schiff and Mannich base of isatin derivatives. Journal of Saudi Chemical Society. Volume 17, Issue 3, July 2013, Pages 337-344.

[89] Karrouchi K et al. Synthesis, antioxidant and analgesic activities of Schiff bases of 4-amino-1, 2, 4-triazole derivatives containing a pyrazole moiety. Annales Pharmaceutiques Françaises. 2016;**74**(6):421-438

[90] Sondhi SM, Singh N, Kumar A, Lozach O, Meijer L. (2006). Synthesis, anti-inflammatory, analgesic and kinase (CDK-1, CDK-5 and GSK-3) inhibition activity evaluation of benzimidazole/ benzoxazole derivatives and some Schiff's bases. Bioorganic & Medicinal Chemistry. 2006;**14**(11):3758-3765

[91] Murtaza S, Akhtar MS, Kanwal F, Abbas A, Ashiq S, Shamim S. Synthesis and biological evaluation of schiff bases of 4-aminophenazone as an anti-inflammatory, analgesic and antipyretic agent. Journal of Saudi Chemical Society. 2017;**21**:S359-S372

[92] Uddin N, Rashid F, Ali S, Tirmizi SA, Ahmad I, Zaib S, et al. Synthesis, characterization, and anticancer activity of Schiff bases. Journal of Biomolecular Structure and Dynamics. 2019;**2019**:1-14

[93] Abd-Elzaher MM, Labib AA, Mousa HA, Moustafa SA, Ali MM, El-Rashedy AA. (2016). Synthesis, anticancer activity and molecular docking study of Schiff base complexes containing thiazole moiety. Beni-Suef University Journal of Basic and Applied Sciences. 2016;**5**(1):85-96

[94] Tadele KT, Tsega TW. Schiff bases and their metal complexes as potential anticancer candidates: A review of recent works. Anti-Cancer Agents in Medicinal Chemistry. 2019;**19**(15):1786-1795

[95] Iacopetta D, Ceramella J, Catalano A, Saturnino C, Bonomo MG, Franchini C, et al. Schiff bases: Interesting scaffolds with promising antitumoral properties. Applied Sciences. 2021;**11**:1877

[96] Thomas AB, Nanda RK, Kothapalli LK, Hamane SC. Synthesis and biological evaluation of Schiff's bases and 2-azetidinones of isonocotinyl hydrazone as potential antidepressant and nootropic agents. Arabian Journal of Chemistry. 2016;**9**(1):S79-S90

[97] Yuldasheva N, Acikyildiz N, Akyuz M, Dambagi LY, Aydin T. The synthesis of Schiff bases and new secondary amine derivatives of p-vanillin and evaluation

of their neuroprotective, antidiabetic, antidepressant and antioxidant potentials. Journal of Molecular Structure. 2022;**1270**:133883

[98] Abdel-Wahab B, Mohamed S, Amr AG, et al. Synthesis and reactions of thiosemicarbazides, triazoles, and Schiff bases as antihypertensive α-blocking agents. Monatshefte fuer Chemie. 2008; **139**:1083-1090

[99] Gopi C, Sastry VG, Dhanaraju MD. Design, synthesis, spectroscopic characterization and anti-psychotic investigation of some novel Azo dye/ Schiff base/chalcone derivatives. Egyptian Journal of Basic and Applied Sciences. 2017;**4**(4):270-287

[100] Udhayakumari D, Inbaraj V. A review on Schiff Base fluorescent Chemosensors for cell imaging applications. Journal of Fluorescence. 2020;**30**:1203-1223

[101] Liu H, Ding S, Lu Q, Jian Y, Wei G, Yuan A. A versatile Schiff Base Chemosensor for the determination of trace Co^{2+}, Ni^{2+}, Cu^{2+}, and Zn^{2+} in the water and its bioimaging applications. ACS Omega. 2022;**7**:7585-7594

[102] Liu H, Liu T, Li J, Zhang Y, Li J, Song J, et al. A simple Schiff base as dual-responsive fluorescent sensor for bioimaging recognition Zn2+ and Al3+ in living cells. Journal of Materials Chemistry B. 2018;**6**:5435-5442

[103] Reyes JB, Flores BMM, Treviño AG, Suárez MAT, Hernández DP, Schott E, et al. Novel fluorescent Schiff bases as Al^{3+} sensors with high selectivity and sensitivity, and their bioimaging applications. Materials Chemistry and Physics. 2019;**233**:89-101

[104] Reyes JCB et al. Two-photon detection of organotin Schiff Base complexes in Cancer cells. Chemistry Selected. 2020;**5**(5):1623-1627

[105] Zhang J, Zhu X, Zhong A, Jia W, Wu F, Li D, et al. New platinum (II) one-armed Schiff base complexes for blue and orange PHOLEDs applications. Organic Electronics. 2017;**42**:153e162

[106] Gusev AN, Kiskin MA, Braga EV, Chapran M, Salyga GW, Baryshnikov GV, et al. Novel zinc complex with an ethylenediamine Schiff Base for high-luminance blue fluorescent OLED applications. Journal of Physical Chemistry C. 2019;**123**(18):11850-11859

[107] Nayak PHA, Naik HSB, Viswanath R, Kirthan BR. Green light emitting fluorescent [Zn(II)(Schiff base)] complexes as electroluminescent material in organic light emitting diodes. Journal of Physics and Chemistry of Solids. 2021;**159**:110288

[108] Berhanu AL, Gaurav MI, Malik AK, Aulakh JS, Kumar V, Kim KH. A review of the applications of Schiff bases as optical chemical sensors. Trends in Analytical Chemistry. 2019;**116**:74-91

[109] Xu J, Liu Y, Hsu SH. Hydrogels based on Schiff Base linkages for biomedical applications. Molecules. 2019;**24**:3005

[110] Liu J, Li J, Yu F, Zhao YZ, Mo XM, Pan JF. In situ forming hydrogel of natural polysaccharides through Schiff base reaction for soft tissue adhesive and hemostasis. International Journal of Biological Macromolecules. 2020;**147**: 653-666

[111] Oiyea EN, Ribeiroa MFM, Katayamaa JMT, Tadinia MC, Balbinoa MA, Eleoterio IC, et al. Electrochemical sensors containing Schiff bases and their transition metal complexes to detect analytes of forensic, pharmaceutical and

environmental interest. A review. Critical Review in Analytical Chemistry. 2019;**49**(6):488-509

[112] Borisov SM, Pommer R, Svec J, Peters S, Novakova V, Klimant I. New red-emitting Schiff base chelates: Promising dyes for sensing and imaging of temperature and oxygen *via* phosphorescence decay time. Journal of Materials Chemistry C. 2018;**6**:8999-9009

[113] Shakour MA, Said WA, Abdellah IM, · Su R, Shafei AE. Low-cost Schiff bases chromophores as efficient co-sensitizers for MH-13 in dye-sensitized solar cells. Journal of Materials Science: Materials in Electronics. 2019;**30**: 5081-5091

[114] Jain P, Pandey G, Kumar D, Chandra S. Prospects of biologically active Schiff's base ligand and metal complexes in drug discovery. Advanced Science Engineering Medicine. 2019; **11**:144-154

[115] Chen G, Fang Y, Zhao X, Tat T, Chen J. Textiles for learning tactile interactions. Nature Electronics. 2021;**4**:175

[116] Basudeb D, Shibashis H. Schiff base compounds as fluorimetric pH sensor: A review. Analytical Methods. 2022;**22**:1-15

Chapter 7

Cytotoxic Activity of Schiff Bases and Their Complexes

Narjala Rama Jyothi and Snehalatha Pulivarthi

Abstract

Various organic chelating agents have many applications in treating the several diseases and they act as antibacterial, antiviral, antimalarial and cytotoxic agents. Among the organic chelating agents thiosemicarbazones and their derivatives play a unique role in various fields of medicine. Thiosemicarbazones and their derivatives find a numerous applications and among them cytotoxic activity occupies a major portion due to the severity of the cancer treatment. In this present chapter we described and discussed the cytotoxic activity of thiosemicarbazones, their derivatives and various metal complexes of them. For this purpose, we reviewed the research articles published in various reputed international journals since 35 years. We summarized the results of those research findings and it is found that among the various metal ions, mostly the platinum and palladium complexes are effective cytotoxic agents than other metal complexes.

Keywords: thiosemicarbazones, metal complexes, cytotoxic activity, Schiff bases

1. Introduction

A Schiff base is a nitrogen analog of a carbonyl compound (aldehyde or ketone) in which the C=O group is replaced by C=N–R group. Generally, Schiff bases are considered as imines bearing a hydrocarbyl group on the nitrogen atom $R_2C{=}NR'$ ($R' \neq H$). Schiff bases are usually synthesized by the condensation of a carbonyl compound with a primary amine as shown in the following **Figure 1**.

Schiff bases had various applications in different fields, such as medical [1], pharmaceutical [2] and biological [3]. Due to the presence of electron donating groups, such as sulfur and nitrogen atoms they can bind with metal ions in enzymes. Among various Schiff bases, thiosemicarbazones occupy a major role by having electron donating nitrogen and sulfur atoms. Thiosemicarbazones are a group of compounds obtained by condensing thiosemicarbazide with carbonyl compounds in the presence of a few drops of glacial acetic acid. These reagents function as good chelating agents and form complexes with several metal ions, by bonding through thionate sulfur atom and hydrazino nitrogen atom. In the last few years, much interest has been directed towards the use of chelating ligands containing sulfur and nitrogen in analytical studies as well as in structural studies of metal complexes. The wide applications and rapid growth in the popularity of sulfur ligands is due to their remarkable property as potential donors to form stable as well as characterized complexes in which the back

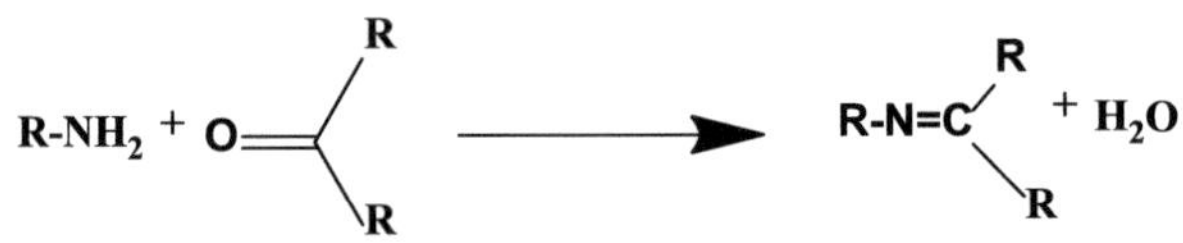

Figure 1.
Formation of Schiff base.

bonding from the metal ion is possible under favorable conditions. In addition, the presence of nitrogen along with sulfur tends to lower the solubility of the complexes, making the isolation of these complexes easier.

Thiosemicarbazones are having great biological activities due to their ability to coordinate to the metal centers in enzymes. A number of studies reveals the biological and pharmacological activities of thiosemicarbazones and their metal complexes, such as anti-bacterial, anti-viral, anti-malarial and antineoplastic [4]. Anticancer activities of thiosemicarbazones were reported by various authors over worldwide [5]. Because of having various applications, in recent years a large number of authors reported the synthesis and characterization studies of different thiosemicarbazone ligands. The importance of thiosemicarbazones both in analytical and biological fields owe to us to synthesize new thiosemicarbazones.

Based on the starting compounds thiosemicarbazones can be classified into four types as shown below:

a. Aldehydes or substituted aldehydes with thiosemicarbazide.

b. Aldehydes or substituted aldehydes with substituted thiosemicarbazide.

c. Ketones or substituted ketones with thiosemicarbazide, and

d. Ketones or substituted ketones with substituted thiosemicarbazide.

Depending on the type of parent aldehyde or ketone used for condensation they can act as unidentate, bidentate or multidentate chelating agents during complexation with metal ions. Thio- and/or phenylthiosemicarbazones are synthesized by the condensation of carbonyl compound (with or without substitutes) with thio (phenylthio-) semicarbazides.

2. Cytotoxic activity of metal thiosemicarbazone complexes

2.1 Pd (II) and Pt(II) complexes

Recently many authors were reported the anticancer activity of thiosemicarbazones and their palladium and platinum complexes. Nguyena et al. [6] reported the anticancer activities of Pd(II) and Pt(II) complexes of 1-picolinoyl-4-substituted Thiosemicarbazones against MCF7 and HepG2 cancer cell lines. Anti-proliferative activity of N-Substituted Indole thiosemicarbazones and its Pd(II) complexes against HeLa S3 cancer cell lines was reported by [7]. Nyawade et al. [8] reported cytotoxic activity of 2-acetyl-5-methyl thiophene and cinnamaldehyde thiosemicarbazones and their palladium(II) complexes cytotoxic activity on human cancer cell lines, Caco-2(Colon), HeLa(Cervical), Hept-G2 (hepatocellular) and PC-3 (Prostate) cell lines. Complexes shows excellent activity than ligands. Matesanz et al. [9] studied anticancer activity of Pd(II) and Pt(II) complexes of pyrrol-2-carbaldehyde N-p-chlorophenylthiosemicarbazone.

The palladium complexes of N-Substituted isatin thiosemicarbazones shows significant *in vitro* cytotoxicity against human breast (MCF) and lung (A549) cell lines [10] cancer cell lines (A549) than the ligand and cisplatin [11]. Pd(II) and Pt(II) complexes of Chlorophenyl substituted thiosemicarbazones were studied and performed the in-vitro anti-cancer activity of the ligand and complexes against T-47D, A2780 and A2780cus R human cell lines. The results indicate that the ligand has more activity than the complexes [12].

Isatin thiosemicarbazone derivatives and its complexes of Pt(II) shows potent activity against human colorectal carcinoma cell line (HCT 116) [13]. 2,6-Diacetylpyridine bis(^{4}N-tolylthiosemicarbazone) complexes of Pd(II) and Pt(II) show the activity against cisplatin resistant A2780cisR tumor cells. These were shows high antiproliferative activity against breast cancer cell MCR-F-7 cells [14]. 2-Oxo-1,2-dihydroquinoline-3-carbaldehyde thiosemicarbazones complexes of Pd(II) shows a good cytotoxic activity against human cancer cell lines such as HeLa, KEp-2, Hep G2, and A431. It was evaluated and found the relationship between the structure and the activity of palladium complexes [15].

6-Methoxy-2-oxo-1,2-dihydro quinoline-3-carbaldehyde ^{4}N-substituted thiosemicarbazones and its complexes of Pd(II) shows better cytotoxic activity against human lung cancer cell line A549 than other complexes and even cisplatin. All the complexes had strong anti-oxidant property [16]. Pd(II), Ni(II), Pt(II) complexes of 5-Acetylbarbituric-^{4}N-dimethylthiosemicarbazones studies show that one of the complex had sufficient cytotoxicity against HeLa cells [17]. 3,5-Diacetyl-1,2,4-triazol bis(^{4}N-substituted thiosemicarbazone) and its complexes of Pd(II) were tested for antiproliferative activity against NCI-H460, A2780 and A2780cisR human cancer cell lines and found that they exhibit low toxicity on kidney cells with respect to cisplatin [18].

Pt(II) complexes of Bis(thiosemicarbazones) of the 3,5-diacetyl-1,2,4-triazol series were tested cytotoxic activity against NCI-H460, A2780, and A2780cisR cancer cell lines and the results indicates that they were active against those cell lines and had high activity against NCI-H460 cell line [19]. 5-Substitutedthiophene-2-Carboxaldehyde thiosemicarbazones and its complexes of Pt(II), Pd(II) shows better cytotoxic activity than free ligand, but palladium complexes shows lesser activity than free ligands [20]. Pt(II) complexes of 3,5-Diacetyl-1,2,4-triazol bis(4,4-dimethylthiosemicarbazone) Anti-proliferative activity of ligand and its complexes was tested against NCI-H460, A2780 and A2780cisR human cancer cell lines. The results show that the compounds exhibits better activity against A2780cisR cell line than cisplatin [21].

2-Acetylpyridine-*N*(4)-methyl-thiosemicarbazone and 2-Acetylpyridine-*N*(4)-phenyl-thiosemicarbazone and its palladium complexes were act as cytotoxic agents similar to cisplatin. All the compounds show better activity against *Mycobacterium tuberculosis* [22]. Cytotoxic activity of the ligands (3,5-Diacetyl-1,2,4-triazol bis(4-substituted) thiosemicarbazones) and their Pd(II) and Pt(II) complexes was tested against human A2780 and A2780cisR epithelial ovarian carcinoma cell lines. The results show that the activity of these compounds were in the range similar to cisplatin drug [23]. Anti-proliferative activity was studied on Phenanthrenequinone thiosemicarbazone and its complexes of Pd(II). The palladium complex had cytotoxic activity against breast cancer cell lines and nontoxic towards the normal mammary epithelial cells [24].

2-Acetyl Pyridine N(4)-Ethyl-thiosemicarbazones, 2-Acetyl Pyridine N(4)-1-(2-pyridyl)-piperazinyl TSC, 2-Formyl Pyridine N(4)-1-(2-pyridyl)-piperazinyl TSC and its complexes of Pt(II), Pd(II) shows antiproliferative activity against gram +ve bacteria but not on gram –ve bacteria. Some of the complexes overcome the cisplatin resistance of A2780/Cp8 cells [25]. 2-Acetylpridine thiosemicarbazone and Platinum complexes were effective on gram + bacteria. They were also effective on yeast. Few of the complexes were exhibits antitumor activity [26]. 2-Acetylpyridine and pyridine-2-carbaldehyde N(4)-ethyl thiosemicarbazones and Platinum complexes were found to be overcome the cisplatin resistant tumor cells, A2780/Cp8 [27].

Pyridine-2-carbaldehyde thiosemicarbazone and its complexes of Pd(II) and Pt(II) exhibits the higher in vivo antitumor activity [28]. Pd(II) and its complexes of 3,5-Diacyl-1,2,4-triazole bis(thiosemicarbazone); 2,6-diacylpyridine bis(TSC); benzyl bis(TSC) shows better antitumor activity against several human, monkey and murine cell lines [29]. *p*-Isopropylbenzaldehyde thiosemicarbazone and its Pd(II) and Pt(II) complexes show cytotoxic activity of the complexes reported against cisplatin resistant tumor cell lines (**Figure 2**) [30].

Isopropylbenzaldehyde thiosemicarbazone and its complexes of Pd(II) and Pt(II) show anticancer activity against several human and murine cell lines were reported by [31]. Phenylacetaldehyde thiosemicarbazone and its complexes of Pd(II) and Pt(II) cytotoxic activity was studied and observed that Cis-DDP-resistant tumor cells has high activity [32].

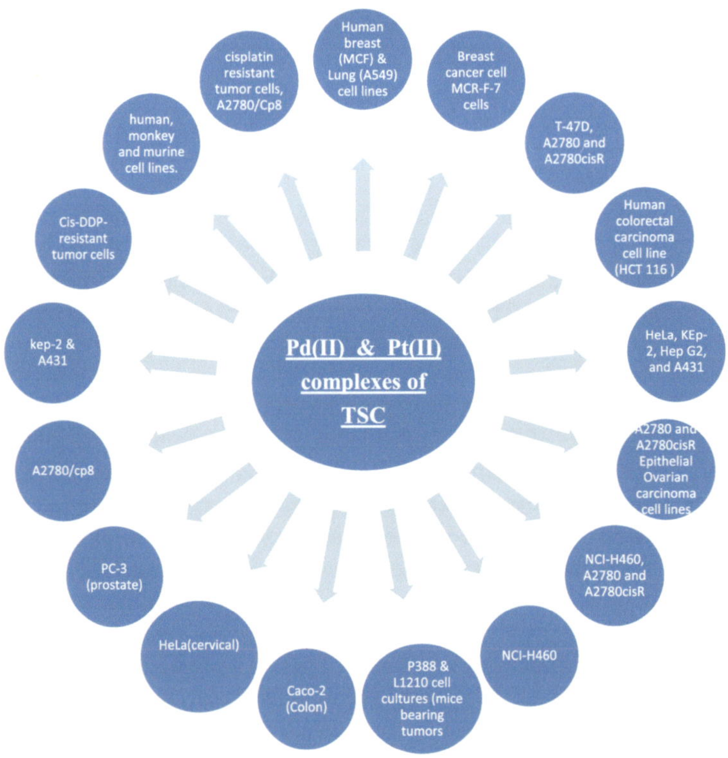

Figure 2.
Cytotoxic activity of Pd(II) and Pt(II) complexes of thiosemicarbazones.

3,5-Diacyl-1,2,4-triazole bis(thiosemicarbazone); 2,6-diacylpyridine bis(TSC); benzyl bis(TSC) and its Pd(II) complexes shows better antitumor activity against several human, monkey and murine cell lines [29]. *p*-Isopropylbenzaldehyde thiosemicarbazone and its complexes of Pd(II) Pt(II) Cytotoxic activity reported against cisplatin resistant tumor cell lines [30]. *p*-Isopropylbenzaldehyde thiosemicarbazone and its complexes of Pd(II) and Pt(II) show anticancer activity against several human and murine cell lines [31].

Cytotoxic activity of phenylacetaldehyde thiosemicarbazone and its Pd(II) and Pt(II) complexes were studied and it was reported that cis-DDP-resistant tumor cells has reacted highly [32]. 2-Acetylpyridine N(4)-methyl, N(4)-ethyl and N(4)-phenyl thiosemicarbazones and its complexes of Pd(II) Antitumor studies indicates that all the palladium complexes were active in the inhibition of DNA synthesis on P388 and L1210 cell cultures (mice bearing tumors) [33]. 2-Acetylpyridine N(4)-propyl, N(4)-dipropyl- and 3-hexamethyleneiminyl thiosemicarbazones and its Palladium complexes does not had antifungal activity against the tested species. But they had significant anti-tumor activity against P388 and L1212 cell cultures [34].

2.2 Copper complexes

Thiosemicarbazone and its copper (II) complexes exhibits high anticancer activity due its highest stability and membrane permeability [35]. 2-Picoline and 5,5-dimethylbipyridine and its Cu(II) complexes cytotoxic activity was reported on MDA-MB-231 breast cancer cell line [36]. Chitosan-functionalized pyridine-based Thiosemicarbazones and their Cu(II) complexes shows antiproliferative activity against MCDK and MCF-7 cancer cell lines and their complexes shows high cytotoxic activity than ligands [37]. *N*-substitution in isatin thiosemicarbazones and its Cu(II) complexes MTT assays were done on A549,HeLa S3, Jurkat and IMR90 cells, its complexes exhibits more cytotoxic activity on HeLa S3 and Jurkat cell lines than cisplatin and morphological changes were observed [38]. Series of thiosemicarbazones and their Cu(II) complexes cytotoxic activity was studied and was noted that ligands chelation with Cu(II) enhances its antitumor activity well [39]. Copper(II) complexes of pyridoxal dithiocarbazate and thiosemicarbazone ligands are prone to study antitumor activities (Ehrlich and S-180 cells) and it was observed that these complexes were very active towards the cell death and they will act as excellent drug against cancer in future [40]. Copper(II) complexes bearing 2-hydroxynaphthaldehyde-based thiosemicarbazones shows cytotoxic activity against lung cancer (A549) cell line and human kidney (HEK-293 T) cell lines and it was observed that these complexes are active in cell death and it was showing excellent medical properties [41].

6-Methyl-2-oxo-quinoline-3-carbaldehydethiosemicarbazone and its Cu(II) complexes Cytotoxic activities were evaluated for both the ligand and three complexes. In vitro anti-tumor studies revealed that copper complex shows better activity towards SK-OV-3 and MGC80-3 tumor cell lines than the commercial anticancer drug, cisplatin. But all the complexes show lower activity against human liver cell lines than cisplatin [42]. 3-Phenyl (substituted)-1-pyridin-2-ylprop-2-en-1-one thiosemicarbazone and its Cu(II) Cytotoxic activity were tested against human cancer cell lines such as HL60, MDA-MB 231, and HCT-116. The results indicate that coordination of copper increases the cytotoxic activity of the compounds [43].

Glyoxal-bis(4-methyl-4-phenyl-3-thiosemicarbazone) and its Copper complex has better cytotoxic activity against various human cancer cell lines than the

Adriamycin, a commercial drug. The copper complex significantly inhibits the growth of tumor HCT 116 xenografts in nude mice [44]. Cu(II) and its complexes of 2-Oxo-1,2-dihydroquinoline-3-carbaldehyde4(*N,N*)-dimethylthiosemicarbazone Cytotoxic activity of the complex and ligand was tested against HeLa, Hep G2, and HEp-2 cancer cell lines and found that the copper complex has significant activity on HeLa cell line than others [45]. N-substituted-2-oxo-1,2-dihydroquinoline-3-carbaldehyde thiosemicarbazones and its complexes of Cu(II) antiproliferative activities were studied the substituent at terminal nitrogen atom was observed. The ethyl and phenyl substituted complexes shows better activity against NIH 3 T3 and HeLa cell lines [46].

α-Heterocyclic-N4-substituted thiosemicarbazones and its Cu(II) complex Anti-proliferative activity were tested against breast cancer cell line SK-BR-3. The ligands shows better catalytic inhibition property of topoisomerase-Iiα than complexes [47]. Cu(II) complexes of 2-Acetylpyridine-4,4-dimethyl-3-thiosemicarbazone, di-2-pyridyl ketone-4,4-dimethyl-3-TSC shows better anti-proliferative activity than ligand [48]. 2-Hydroxy-8-R-tricyclo [7.3.1.0.^{2,7}] tridecane-13-one thiosemicarbazone and its complexes of Cu(II), Pd(II) shows Anti-microbial activities and cytotoxic activities of the compounds were reported [49]. Salicylaldehyde semi–/thiosemicarbazones and its complexes of Cu(II) were tested for cytotoxic activity against MCF-7 human breast cancer cell lines. The results revealed that ligands were inactive but copper complex of thiosemicarbazone was more active than others [50].

5-Formyluracil thiosemicarbazone derivatives and its Cu(II) Complexes were exhibits DNA interaction by electrostatic and groove binding. But these were no activity against human leukemic cell line U937 [51]. α-Ketoglutaric acid thiosemicarbazone and its Cu(II) Copper complexes has antiproliferative activity against human cell line U937 and no effect on the K562 cell line [52]. 10-Deacetylbaccatin thiosemicarbazone and its Cu(II) Cytotoxic activity of the ligand and its complex was tested against human breast cancer cell line MCF-7 [53].

Ref. [54] reported the cytotoxic activity of Cu(II) and its complexes of 2-Acetylpyrazone-N-substituted thiosemicarbazones. Ref. [55] studied the anticancer activity of 5-Formyluracil thiosemicarbazone and its Cu(II) complexes against human leukemic cell lines K562 and CEM (**Figure 3**).

2.3 Cu(II) and Zn(II) complexes

Hydroxyquinoline-thiosemicarbazones and its Cu(II) and Zn(II) complexes shows more cytotoxic activity towards human lung cell lines (A549) than their parent ligands [56]. Methyl pyruvate thiosemicarbazones and its Cu(II), Zn(II) complexes anti-proliferative activity studies reveals that copper complex was most effective on human leukemic cell line U937 than other compounds [57].

2.4 Cu(II) and Ni(II) complexes

Cu(II), Ni(II) complexes and its N-Ethyl-2-(phenyl(pyridin-2-yl)methylene) hydrazine carbothioamide cytotoxic activity was evaluated against human lung cancer cell lines (A549) and normal cell lines (L929). The results indicates that the copper complexes shows better activity than nickel complexes and both complexes were less harmful to normal cells [58]. 2,4′-Dibromoacetophenone thiosemicarbazone and its Cu(II), Ni(II), Pd(II) complexes, Antioxidant and antitumor activities

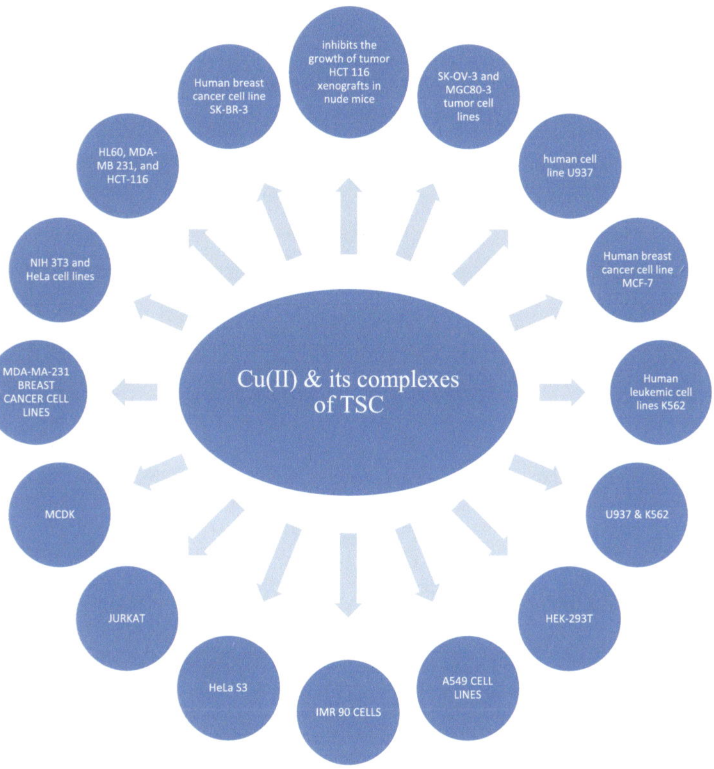

Figure 3.
Cytotoxic activity of Cu(II) complexes of thiosemicarbazones.

were studied. Nickel complex shows good antitumor activity against HepG2 hepatoblastoma cell lines [59]. Cu(II) and Ni(II) complexes and its Cinnamaldehyde and cuminaldehyde thiosemicarbazones were tested in vitro anti-leukemic activity on U937 human cell line. Metal complexes shows better activity than ligands [60]. 3,4-Difluoroacetophenone thiosemicarbazone; 2-bromo-4′-chloroacetophenone TSC and its Cu(II), Ni(II) complexes were tested for antitumor activity against HepG2 human hepatoblastoma cells. The results indicates that the copper complexes of the both ligands had better activity than others against cancer cell line [61]. (Z)-2-(Amino(pyridin-2-yl) methylene)-N-methylhydrazine carbothioamide and its Cu(II), Zn(II) complexes Cytotoxic activity was studied against HeLa, HepG-2 and SGC-7901 cell lines. The results shows that copper complex had better activity among the others [62].

Bis(citronella thiosemicarbazone); Pyridoxal TSC and its Ni(II), Cu(II) Metal complexes were tested for antiretro-viral activity against HIV-1 and HTLV-1/−2. The results indicates that copper complex has potent anti HIV activity (**Figure 4**) [63].

2.5 Cu(II), Ni(II) and Zn(II) complexes

2-Methoxybenzaldehyde-S-2-methylbenzyl dithiocarbamate and 3-methoxybenzalde hyde-S-2-methylbenzyl dithiocarbamate and its Cu(II), Ni(II), Zn(II) complexes were inactive against MCF-7 and MDA-MB-231 breast cancer cell lines [64].

Figure 4.
Cytotoxic activity of Cu(II), Zn(II) and Ni(II) complexes of thiosemicarbazones.

2.6 Ni(II) complexes

Ni(II) and its complexes of 3-Methoxy-salicylaldehyde-4(N)-substituted thiosemicarbazones exhibits strong anti-oxidant property with strong radical scavenging ability. Some of the complexes shows better anticancer activity against lung cancer cell lines (A549) than the ligand and cisplatin [11]. N(4)-substituted thiosemicarbazone and its Ni(II) complexes Anti-cancer activity was tested against human breast cancer cell lines (MCF-7). All the complexes shows moderate activity compared to commercial anti-cancer drug, cisplatin [65].

Ni(II) and its complexes of dinucleating bis(thiosemicarbazones) were tested for cytotoxic activity against human cancer cell lines (A549 and HepG2). One of the nickel complex shows better activity against A549 cell line than cisplatin drug [66].

2-Hydroxy-1-naphthaldehydethiosemicarbazone; salicylaldehyde-4(N)-ethylthiosemicarbazone; 2-hydroxy-1-naphtha ldehyde-4(N)-ethyl TSC and Ni(II) complexes were tested for cytotoxic activity against human cancer cell lines (A549 and HepG2). One of the nickel complex shows better activity against A549 cell line than cisplatin drug [67]. Ortho-Naphthaquinone thiosemicarbazone; ortho-Naphthaquinone semicarbazone Anti-cancer activity and its Ni(II) complexes were tested against MCF-7 human breast cancer cell lines and results revealed that semicarbazone and nickel complexes were more active than thiosemicarbazone ligand [68].

p-Fluorobenzaldehdye thiosemicarbazones and its nickel complexes shows remarkable inhibition activity against human leukemic cell line U937 (**Figure 5**) [69].

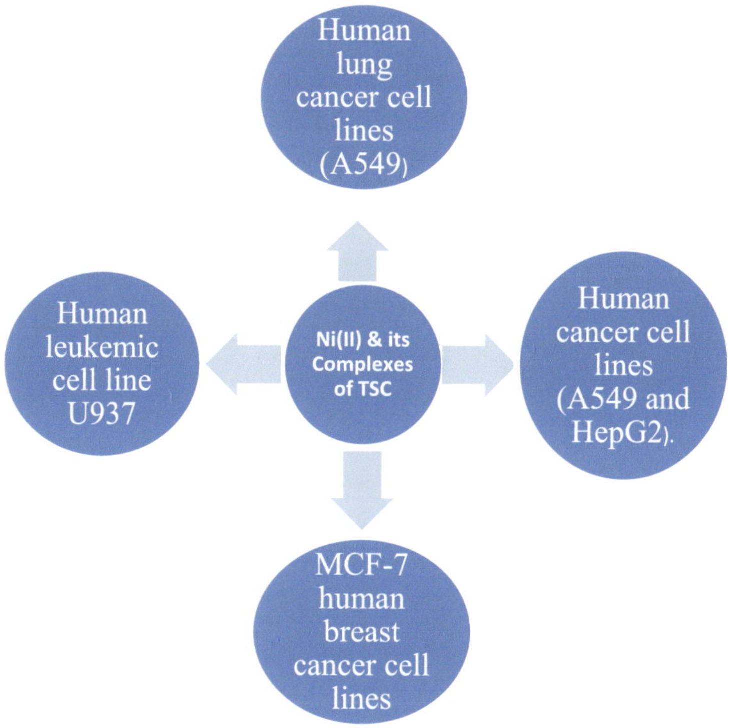

Figure 5.
Cytotoxic activity of Ni(II) complexes of thiosemicarbazones.

2.7 Fe(II) and Fe(III) complexes

α-N-heterocyclic thiosemicarbazone and its complex of Fe(III) Anti-cancer activity was studied against human breast cancer cell lines, cervical cancer cell lines and liver cell lines. The results show that iron complex has better activity than its ligand. This study also evaluates the anti-cancer mechanism [70]. 2-Acetylpyridine thiosemicarbazones and its Fe(III) complexes Antiproliferative property were reported. The structure and activity relationship was determined [71]. 2-Acetylpyridine *N,N*-dimethylthiosemicarbazone; 2-acetyl pyridine *N*-pyrrolidinyl TSC; acetyl pyrazine *N,N*-dimethyl TSC; acetyl pyrazine *N*-pyrrolidinyl and acetylpyrazine-N-piperidinyl TSC and its Ga(III), Fe(III) complexes Cytotoxic activity were studied and it was reported that ligands were enhanced by the chelation with Ga(III) while weakens with Fe(III) [72].

Iron and its complexes of Series of Di-2-pyridyl ketone thiosemicarbazones had good antiproliferative activity against the tested tumor cells [73]. 1-Formylisoquinoline thiosemicarbazone; 4-Methyl-5-amino-1-formylisoquinoline TSC and its Fe(II), Fe(III) complexes were active against P 388 lymphocytic leukemia test system in mice (**Figure 6**) [74].

2.8 Organo Sn complexes

Cytotoxic activity of 2-Hydroxy-5-methoxy benzaldehyde-N(4)-methylthiosemicarbazone and its complexes Organotin (IV) were tested against human colorectal (HCT 116) cell lines and the results show that the complexes had better activity than

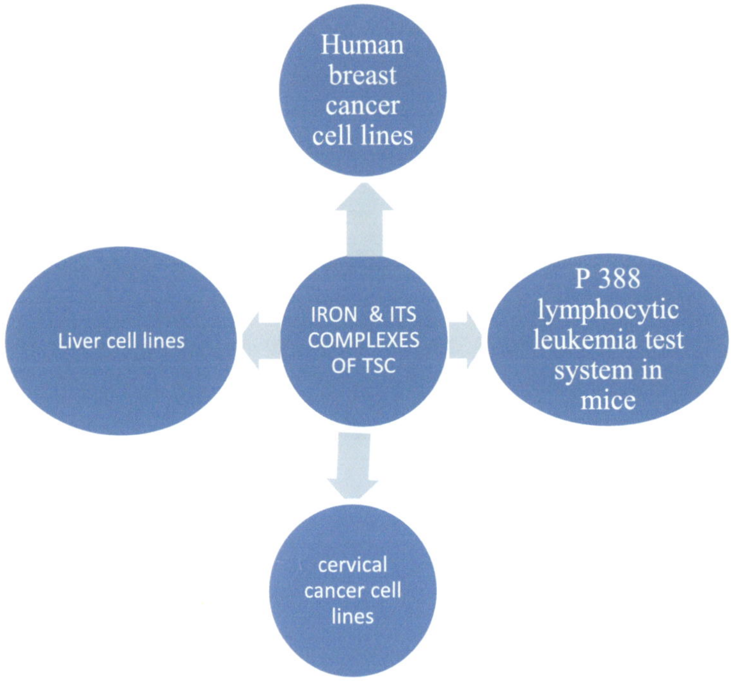

Figure 6.
Cytotoxic activity of iron complexes of thiosemicarbazones.

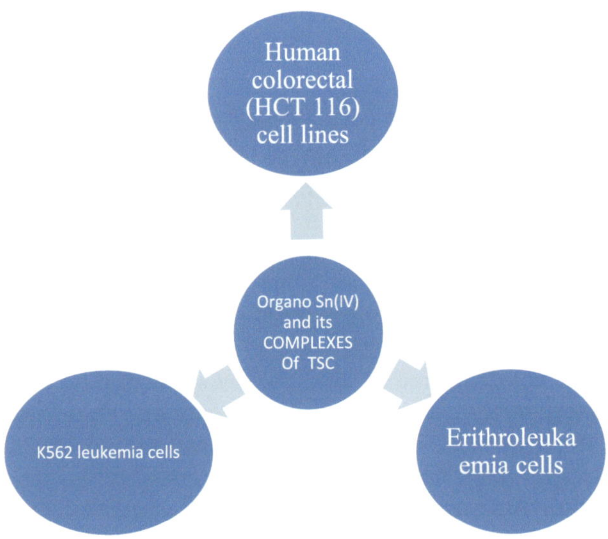

Figure 7.
Cytotoxic activity of Organo tin complexes of thiosemicarbazones.

the ligand [75]. 2-Benzoylpyridine N(4)-phenyl thiosemicarbazone; 2-Acetyl pyrazine N(4)-phenyl TSC and its Diorgano Sn(IV), Antibacterial activity and cytotoxic activity against K562 leukemia cells of the free ligands and complexes were reported [76]. Pyridoxal thiosemicarbazone and its complexes of Diorgano Sn(IV) antitumor activity were studied and observed that Ethyl, butyl, and phenyl substituted compounds suppress the proliferation of friend erythroleukemia cells (**Figure** 7) [77].

3. Miscellaneous

2-Benzoylpyridine N(4)-cyclohexyl thiosemicarbazone-Anti-proliferative activity of the ligand and its indium complexes was tested against human hepatocellular carcinoma. The studies revealed that the indium complex has better activity than the others [78]. N4-(2-Hydroxy-5-chlorobenzyliden e)-2-amino-5-chlorobenzophenone thiosemicarbazone; N4-(2-Hydroxy naphthalene-1-carbaldehyde)-2-amino-5-chlorobenzophenone TSC and its complexes of Ru(II) cytotoxic activity was studied and reported that, all the complexes showed better in vitro cytotoxic activity against MCF-7, Hop62, MDA-MB cell lines [79]. 7-Chloroquinoline thiosemicarbazone and its complexes of Ga(III), Cytotoxic and antimalarial activity were tested and proved that the complex shows 31 times better activity on colon cancer cell line than etoposide. The complex has better antimalarial activity against *Plasmodium falciparum* than lumefantrine [80].

2-Acetylpyridine-N(4)-Orthochlorophenyl thiosemicarbazone and their complexes of Ga(III), Sn(IV), Pd(II) and Pt(II) Cytotoxic activity was reported. The results indicate the ligands and metal complexes showed better cytotoxic activity [81]. (Z)-(2-((1,3-Diphenyl-1H-pyrazol-4-yl) methylene) and hydrazinyl) (pyridin-2-ylamino) methane-thiol thiosemicarbazones and their complexes of Mn(II), Co(II), Ni(II), Cu(II), Zn(II), Cd(II) Cd(II) and Zn(II) complexes had strong antioxidative property. Ni(II) complex shows excellent activity against HepG2 and MCF-7 cancer cell line [82]. Cd(II) and its complexes of 2-Acetylyuridine-N4-substituted thiosemicarbazones and their Cytotoxic activity against human breast cancer cell lines was tested and found that one of cadmium complexes has better activity than the cisplatin drug [83]. Bi(III) and its complexes of 2,6-Diacetylpyridine bis (4N-methylthiosemicarbazone) have much more anti-bacterial and anticancer activity than its parent ligand against *Bacillus cereus* and *Salmonella Typhimurium* and K562 leukemia cells [84]. Ru(II) and its complexes of Benzaldehyde thiosemicarbazones and their Cytotoxic activity were evaluated against the WHCO1 esophageal cancer cell line. Among the tested complexes two had moderate activity against the cancer cell line [85]. 2-Acetylpyridine thiosemicarbazone; 2-Acetylpyridine N(4)-methyl TSC; 2-Benzoylpyridine TSC; 2-Benzoyl pyridine N(4)-methyl TSC; 2-Benzoylpyridine N(4)-phenyl TSC and its complexes of Mn(II), Co(II), Zn(II) was tested against its cytotoxicity and reported that compounds show significant activity against K562 leucocythemia cancel cell line [86].

2-Pyridineformamide thiosemicarbazones, 2-formyl and 2-acetyl acetyl pyridine thiosemicarbazones and its complexes of Ga(III) antiproliferative activity were evaluated against human cancer cell lines (MCF-7, T24, A549 and mouse L-929). 2-acetyl pyridine thiosemicarbazone shows higher activity than others against all cancer cell lines [87]. Kowol et al. [72] studied the anticancer activity of 2-Acetylpyridine N, N-dimethyl TSC, 2-acetyl pyridine N-pyrrolidinyl TSC, acetyl pyrazine N, N-dimethyl TSC, acetyl pyrazine N-pyrrolidinyl and acetyl pyrazine-N-piperidinyl TSC and its complexes of Ga(III), Fe(III). Cytotoxic activity of the ligands was enhanced by the chelation with Ga(III) while weakening with Fe(III). Noblia et al. [88] reported cytotoxic activity of 5-Bromo salicylaldehyde semicarbazone; 2-Hydroxy-naphtalen-1-carboxaldehyde semicarbazone and their vanadium complexes. The results indicate that the complexes had selective activity against TK-10 cell line. Anticancer activity of ortho-Naphthaquinone thiosemicarbazone and its Cu(II),Ni(II), Pd(II) and Pt(II) complexes were tested against MCF7 human breast cancer cell lines and results revealed that chelation of metal ion into ligand was enhanced its activity [89]. Casas et al. reported the cytotoxic activity of Formylferrocene thiosemicarbazones and their Au(III) complexes were evaluated against HeLa cell lines and found that antiproliferative activity was similar to cisplatin, a commercial drug.

Arion et al. [90] reported cytotoxic activity of 2-Acetylpyridine 4 N-dimethyl TSC and its Ga(III) complex against human cancer cell lines SW480, SK-BR-3, and 41 M were tested and results revealed that complexes show slightly higher activity than ligand. Antiproliferative activity of Acenaphthenequinone thiosemicarbazone and its Ni(II), Fe(II), Cu(II), Zn(II)were reported and found that Cu(II) complex shows better antiproliferative activity than others. Perez et al. [91] tested the cytotoxic activity of p-Isopropyl benzaldehyde and methyl 2-pyridyl ketone thiosemicarbazones and their Zn(II), Cd(II) against various cell lines. Results indicate that zinc complex was proven as a more potent antitumor agent. Jayasree and Araindakshan [92] reported the antitumor activity of Acetoacetanilide thiosemicarbazone and its Mn(II), Zn(II), Cd(II), Co(II), Fe (IIII) complexes against Ehrlich Ascites tumor cells and found that metal complexes were more active than free ligand. Mohan et al. [93] studied the antitumor activity of 2,6-Diacetylpyridine bis(N4-azacyclic thio semicarbazones) and their Mn(II), Fe(III), Fe(II), Co(II), Ni(II), Cu(II), Zn(II), Pt(II) complexes against P 388 lymphocytic leukemia test system in mice shows that copper complex shows better activity than others.

4. Conclusions

From the above discussion it is concluded that in most of the studies the metal complexes show greater cytotoxic activity than the free ligands. In metal complexes most of the researchers found the activity of palladium and platinum complexes than other metal complexes. There is a gap between structure activity relationship between metal complexes and their activity towards specific species. Metal complexes of thiosemicarbazones shows great antiproliferative activity and destroys cell completely. Several derivatives of TSC and its metal complexes such as Pt(II), Pd(II), Cu(II), Fe(II), Ni(II), etc., shows excellent cytotoxic activity.

Author details

Narjala Rama Jyothi[1]* and Snehalatha Pulivarthi[2]

1 Department of Basic Sciences and Humanities, School of Engineering and Technology, Sri Padmavati Mahila Visvavidyalayam, Tirupati, India

2 Department of Mechanical Engineering, School of Engineering and Technology, Sri Padmavati Mahila Visvavidyalayam, Tirupati, India

*Address all correspondence to: ramadasaradhi@gmail.com

References

[1] Azarkish M, Akbari A, Sedaghat T, Simpson J. Heteroleptic complexes of Zn(II) based on 1-(5-bromo-2-hydroxybenzylidene)-4-phenylthiosemicarbazide: Synthesis, structural characterization, theoretical studies and antibacterial activity. Journal of Molecular Structure. 2017;**1134**:126-134

[2] Guveli S, Ozdemir N, Ulkuseven B, Bal-Demirci T. Divalent nickel complexes of thiosemicarbazone based on 5-bromosalicylaldehyde and triphenylphosphine: Experimental and theoretical characterization. Polyhedron. 2016;**113**:16-24

[3] Matesanz A, Tapia S, Souza P. First 3,5-diacetyl-1,2,4-triazol derived mono(thiosemicarbazone) and its palladium and platinum complexes: Synthesis, structure and biological properties. Inorganica Chimica Acta. 2016;**445**:62-69

[4] Indoria S, Lobana TS, Sood H, Arora DS, Hundal G, Jasinski JP. Synthesis, spectroscopy, structures and antimicrobial activity of mixed-ligand zinc(II) complexes of 5-nitro-salicylaldehyde thiosemicarbazones. New Journal of Chemistry. 2016;**40**:3642-3653

[5] StacyAE, PalanimuthuD, BernhardtPV, KalinowskiDS, JanssonPJ, RichardsonDR. Zinc(II)–Thiosemicarbazone Complexes Are Localized to the Lysosomal Compartment Where They Transmetallate with Copper Ions to Induce Cytotoxicity. Journal of Medicinal Chemistry. 2016;**59**:4965-4984

[6] Nguyena HH, Phama QT, Phung QM, Le CD, Thi Ngoc TT, Pham O, et al. Synthesis structures and biological activities of Pd(II) and Pt(II) complexes with 1-picolinoyl-4-substituted thiosemicarbazones. Journal of Molecular Structure. 2022;**1269**:133871

[7] Balakrishnana N, Haribabu J, Malekshah RE, Swaminathan S, Balachandran C, Bhuvanesh N, et al. Effect of *N*-benzyl group in indole scaffold of thiosemicarbazones on the biological activity of their Pd(II) complexes: DFT, biomolecular interactions, *in silico* docking, ADME and cytotoxicity studies. Inorganica Chemica Acta. 2022;**534**:120805

[8] Nyawade EA, Sibuyi NRS, Meyer M, Lalancette R, Onani MO. Synthesis, characterization and anticancer activity of new 2-acetyl-5-methyl thiophene and cinnamaldehyde thiosemicarbazones and their palladium(II) complexes. Inorganica Chemica Acta. 2021;**515**:120036

[9] Matesanz AI, Jimenez-Faraco E, Ruiz MC, Balsa LM, Ranninger CN, León IE, et al. Mononuclear Pd(II) and Pt(II) complexes with an α-N-heterocyclic thiosemicarbazone: Cytotoxicity, solution behaviour and interaction versus proven models from biological media. Inorganic Chemistry. 2018;**Front,5**:73-83

[10] Muralisankar M, Basheer SM, Haribabu J, Bhuvanesh NSP, Karvembu R, Sreekanth A. An investigation on the DNA/protein binding, DNA cleavage and in vitro anticancer properties of SNO pincer type palladium(II) complexes with N-substituted isatin thiosemicarbazone ligands. Inorganic Chimica Acta. 2017;**466**:61-70

[11] Kalairasi G, Umadevi C, Shanmugapriya A, Kalaivani P, Dallemer F, Prabhakaran R. DNA(CT), protein (BSA) binding studies, anti-oxidant and

cytotoxicity studies of new binuclear Ni(II) complexes containing 4(N)-substituted thiosemicarbazones. Inorganica Chimica Acta. 2016;**453**:547-558

[12] Matesanz AI, Albacete P, Souza P. Synthesis and characterization of new bioactive mono(thiosemicarbazone) ligand based on 3,5-diacetyl-1,2,4-triazone diketone and its palladium and platinum complexes. Polyhedron. 2016;**109**:161-165

[13] Ali AQ, Teoh SG, Salhin A, Eltayeb NE, Khadeer Ahamed MB, Abdul Majid AMS. Synthesis of platinum(II) complexes of isatin thiosemicarbazones derivatives: In Vitro anti-cancer and deoxyribose nucleic acid binding activities. Inorganica Chimica Acta. 2014;**416**:235-244

[14] Matesanz AI, Leitao I, Souza P. Palladium(II) and platinum(II) bis(thiosemicarbazone) complexes of the 2,6-diacetylpyridine series with high cytotoxic activity in cisplatin resistant A2780cisR tumor cells and reduced toxicity. Journal of Inorganic Biochemistry. 2013;**125**:26-31

[15] Ramachandran E, Raja DS, Rath NP, Natarajan K. Role of substitution at terminal nitrogen of 2-Oxo-1,2-dihydroquinoline-3-carbaldehyde thiosemicarbazones on the coordination behavior and structure and biological properties of their palladium(II) complexes. Inorganic Chemistry. 2013;**52**:1504-1514

[16] Ramachandran E, Raja DS, Bhuvanesh NSP, Nataraja K. Mixed ligand palladium (II) complexes of 6-methoxy-2-oxo-1,2-dihydroquinoline-3-carbaldehyde 4N-substituted thiosemicarbazones with triphenylphosphine co-ligand: Synthesis, crystal structure and biological properties. Dalton Transactions. 2012;**41**:13308-13323

[17] Castineiras A, Fernandez-Hermida N, Garcia-Santos I, Gomez-Rodriguez L. Neutral Ni^{II}, Pd^{II} and Pt^{II} ONS-pincer complexes of 5-acetylbarbituric-4Ndimethylthiosemicarbazone: Synthesis, characterization and properties. Dalton Transactions. 2012;**41**:13486-13495

[18] Matesanz AI, Hernandez C, Rodriguez A, Souza P. 3,5-Diacetyl-1,2,4-triazol bis(4N-substituted thiosemicarbazone) palladium(II) complexes: Synthesis, structure, anti-proliferative activity and low toxicity on normal kidney cells. Journal of Inorganic Biochemistry. 2011;**105**:1613-1622

[19] Matesanz AI, Hernandez C, Rodriguez A, Souza P. Novel bis(thiosemicarbazones) of the 3,5-diacetyl-1,2,4-triazol series and their platinum(II) complexes: Chemistry, antiproliferative activity and preliminary nephrotoxicity studies. Dalton Transactions. 2011;**40**:5738-5745

[20] Iyidogan AK, Tasdemir D, Oruc-Emre EE, Balzarini J. Novel platinum(II) and palladium(II) complexes of thiosemicarbazones derived from 5-substitutedthiophene-2-carboxaldehydes and their antiviral and cytotoxic activities. European Journal of Medicinal Chemistry. 2011;**46**:5616-5624

[21] Matesanz AI, Joie C, Souza P. Chemistry, antiproliferative activity and low nephron toxicity of 3,5-diacetyl-1,2,4-triazol bis(4N-thiosemicarbazone) ligands and their platinum(II) complexes. Dalton Transactions. 2010;**39**:7059-7065

[22] Maia PIDS, Graminha A, Pavan FR, Leite CQF, Batista AA, Back DF, et al. Palladium(II) complexes with thiosemicarbazones. Synthesis, characterization and cytotoxicity against breast cancer cells and anti-*Mycobacterium*

tuberculosis activity. Journal of Brazilian Chemical Society. 2010;7:1177-1186

[23] Matesanz AI, Souza P. Palladium and platinum 3,5-diacetyl-1,2,4-triazol bis (thiosemicarbazones): Chemistry, cytotoxic activity and structure-activity relationship. Journal of Inorganic Biochemistry. 2007;**101**:245-253

[24] Padhye S, Afrasiabi Z, Sinn E, Fok J, Mehta K, Rath N. Antitumor metallothiosemicarbazonates: Structure and antitumor activity of palladium complex of phenanthrenequinone thiosemicarbazone. Inorganic Chemistry. 2005;**44**:1154-1156

[25] Kovala-Demertzi D, Demertzis MA, Filiou E, Pantazaki AA, Yadav PN, Miller JR, et al. Platinum(II) and palladium(II) complexes with 2-acetyl pyridine 4N-ethyl thiosemicarbazone able to overcome the *cis*-platin resistance. Structure, antibacterial activity and DNA strand breakage. Biometals. 2003;**16**:411-418

[26] Kovala-Demertzi D, Demertzis MA, Miller JR, Papadopoulou C, Dodorou C, Filousis G. Platinum(II) complexes with 2-acetyl pyridine thiosemicarbazone: Synthesis, crystal structure, spectral properties, antimicrobial and antitumour activity. Journal of Inorganic Biochemistry. 2001;**86**:555-563

[27] Kovala-Demertzi D, Yadav PN, Demertzis MA, Coluccia M. Synthesis, crystal structure, spectral properties and cytotoxic activity of platinum(II) complexes of 2-acetyl pyridine and pyridine-2-carbaldehyde N(4)-ethyl-thiosemicarbazones. Journal of Inorganic Biochemistry. 2000;**78**:347-354

[28] Kovala-Demertzi D, Miller JR, Kourkoumelis N, Hadjikakou SK, Demertzis MA. Palladium(II) and platinum(II) complexes of pyridine-2-carbaldehyde thiosemicarbazone with potential biological activity. Synthesis, structure and spectral properties. Extended network via hydrogen bon linkages of [Pd(PyTsc) Cl]. Polyhedron. 1999;**18**:1005-1013

[29] Matesanz AI, Perez JM, Navarro P, Moreno JM, Colacio E, Souza P. Synthesis and characterization of novel palladium(II) complexes of bis (thiosemicarbazone). Structure, cytotoxic activity and DNA binding of Pd(II)-benzyl bis(thiosemicarbazonate). Journal of Inorganic Biochemistry. 1999;**76**:29-37

[30] Quiroga AG, Perez JM, Lopez-Solera I, Montero EI, Masaguer JR, Alonso C, et al. Binuclear chloro-bridged palladated and platinated complexes derived from p-isopropylbenzaldehyde thiosemicarbazone with cytotoxicity against cisplatin resistant tumor cell lines. Journal of Inorganic Biochemistry. 1998;**69**:275-281

[31] Quiroga AG, Perez JM, Lopez-Solera I, Masaguer JR, Luque A, Roman P, et al. Novel tetranuclear orthometalated complexes of Pd(II) and Pt(II) derived from p-isopropylbenzaldehyde thiosemicarbazone with cytotoxic activity in cis-DDP resistant tumor cell lines. Interaction of these complexes with DNA. Journal of Medicinal Chemistry. 1998;**41**:1399-1408

[32] Quiroga AG, Perez JM, Montero EI, Masaguer JR, Alonso C, Navarro-Ranninger C. Palladated and platinated complexes derived from phenylacetaldehyde thiosemicarbazone with cytotoxic activity in cis-DDP resistant tumor cells. Formation of DNA interstrand cross-links by these complexes. Journal of Inorganic Biochemistry. 1998;**70**:117-123

[33] Kovala-Demertzi D, Domopoulou A, Demertzis MA, Valle G, Papageorgiou A.

Palladium(II) complexes of 2-acetylpyridine N(4)-methyl, N(4)-ethyl and N(4)-phenyl-thio semicarbazones. Crystal structure of chloro(2-acetylpyridine N(4)-methylthiosemicarbazonato) palladium(II). Synthesis, spectral studies, *in vitro* and *in vivo* antitumor activity. Journal of Inorganic Biochemistry. 1997;**68**:147-155

[34] Kovala-Demertzi D, Domopoulou A, Demertzis MA, Papageorgiou A, West DX. Palladium(II) complexes of 2-acetylpyridine N(4)-propyl, N(4)-dipropyl- and 3-hexa methyleneiminyl thiosemicarbazones with potentially interesting biological activity. Synthesis, spectral properties, antifungal and in vitro antitumor activity. Polyhedron. 1997;**16**(20):3625-3633

[35] Pósa V, Hajdu B, Tóth G, Dömötör O, Kowol CR, Bernhard KK, et al. The coordination modes of (thio) semicarbazone copper(II) complexes strongly modulate the solution chemical properties and mechanism of anticancer activity. Journal of Inorganic Biochemistry. 2022;**231**:111786

[36] Mathews NA, Kurup MRP. Copper(II) complexes as novel anticancer drug: Synthesis, spectral studies, crystal structures, *in silico* molecular docking and cytotoxicity. Journal of Molecular Structure. 2022;**1258**:132672

[37] Adhikari HS, Garai A, Manandhar KD, Yadav PN. Pyridine-based NNS tridentate chitosan thiosemicarbazones and their copper(II) complexes: Synthesis, characterization, and anticancer activity. ACS. 2022;**7**(35):30978-30988

[38] Haribabu J, Alajrawy OI, Jeyalakshmi K, Balachandran C, Anantha Krishnan D, Bhuvanesh N, et al. *N*-substitutioninisatinthiosemicarbazones decides nuclearity of Cu(II) complexes – Spectroscopic, molecular docking and cytotoxic studies. Spectrochemica Acta Part A: Molecular and Bimolecular Spectroscopy. 2021;**246**:118963

[39] Pitucha M, Plewko AK, Czylkowska A, Rogalewicz B, Drozd M, Iwan M, et al. Influence of complexation of thiosemicarbazone derivatives with Cu (II) ions on their antitumor activity against melanoma cells. International Journal of Molecular Sciences. 2021, 2021;**22**(6):3104

[40] Gatto CC, Chagas MAS, Lima IJ, Andrade FM, Silva HD, Abrantes GR, et al. Copper(II) complexes with pyridoxal dithiocarbazate and thiosemicarbazone ligands: Crystal structure, spectroscopic analysis and cytotoxic activity. Transition Metal Chemistry. 2019;**44**:329-340

[41] Aneesrahman KN, Rohini G, Bhuvanesh NSP, Sundararaj S, Musthafa M, Sreekanth A. In vitro biomolecular interaction studies and cytotoxic activities of newly synthesised copper(II) complexes bearing 2-hydroxynaphthaldehyde-based thiosemicarbazone. Chemistry Select. 2019;**3**(28), 2018:8118-8130

[42] Zou BQ, Lu X, Qin QP, Bai YX, Zhang Y, Wang M, et al. Three novel transitional metal complexes of 6-methyl-2-oxo-quinoline-3-carbaldehyde thiosemicarbazone: Synthesis, crystal structure, cytotoxicity and mechanism of action. RSC Advances. 2017;7:17923-17933

[43] Da Silva JG, Despaigne AAR, Louro SRW, Bandeira CC, Souza-Fagundes EM, Beraldo H. Cytotoxic activity, albumin and DNA binding of new copper(II) complexes with chalcone-derived thiosemicarbazones. European Journal of Medicinal Chemistry. 2013;**65**:415-426

[44] Palanimuthu D, Shinde SV, Somasundaram K, Samuelson AG. In vitro and in vivo anticancer activity of copper bis(thiosemicarbazone) complexes. Journal of Medicinal Chemistry. 2013;**56**:722-734

[45] Raja DS, Bhuvanesh NSP, Natarajan K. Biological evaluation of a novel water soluble sulphur bridged binuclear copper(II) thiosemicarbazone complex. European Journal of Medicinal Chemistry. 2011;**46**:4584-4594

[46] RajaDS,ParamaguruG,BhuvaneshNSP, Reibenspies JH, Renganathan R, Natarajan K. Effect of terminal N-substitution in 2-oxo-1,2-dihydroquinoline-3-carbaldehyde thiosemicarbazones on the mode of coordination, structure, interaction with protein, radical scavenging and cytotoxic activity of copper(II) complexes. Dalton Transactions. 2011;**40**:4548-4559

[47] Zeglis BM, Divilov V, Lewis JS. Role of metalation in the topoisomerase IIα inhibition and antiproliferation activity of a series of α-heterocyclic-N^4-substituted thiosemicarbazones and their Cu(II) complexes. Journal of Medicinal Chemistry. 2011;**54**:2391-2398

[48] Jansson PJ, Sharpe PC, Bernhardt PV, Richardson DR. Novel thiosemicarbazones of the ApT and DpT series and their copper complexes: Identification of pronounced redox activity and characterization of their antitumor activity. Journal of Medicinal Chemistry. 2010;**53**:5759-5769

[49] Rosu T, Pahontu E, Pasculescu S, Gerogescu R, Stanica N, Curaj A, et al. Synthesis, characterization antibacterial and antiproliferative activity of novel Cu(II) and Pd(II) complexes with 2-hydroxy-8-R-tricyclo[7.3.1.0.^{2,7}] tridecane-13-one thiosemicarbazone. European Journal of Medicinal Chemistry. 2010;**45**:1627-1634

[50] Patole J, Padhye S, Padhye S, Newton CJ, Anson C, Powell AK. Synthesis, characterization and in vitro anti-cancer activities of semicarbazone and thiosemicarbazone derivatives of salicylaldehyde and their complexes against human breast cancer cell line MCF-7. Indian Journal of Chemistry. 2004;**43A**:1654-1658

[51] Baldini M, Belicchi-Ferrari M, Bisceglie F, Pelosi G, Pinelli S, Tarasconi P. Cu(II) complexes with heterocyclic substituted thiosemicarbazones: The case of 5-formyluracil. Synthesis, characterization, X-ray structures, DNA interaction studies, and biological activity. Inorganic Chemistry. 2003;**42**:2049-2055

[52] Belicchi-Ferrari M, Bisceglie F, Fava GG, Pelosi G, Tarasconi P, Albertini R, et al. Synthesis, characterization and biologically activity of two new polymeric copper(II) complexes with α-Ketoglutaric acid thiosemicarbazone. Journal of Inorganic Biochemistry. 2002;**89**:36-44

[53] Murugkar A, Padhye S, Roy SG, Wagh U. Metal complexes of Taxol precursor 1. Synthesis, characterization and antitumor activity of the copper complex of 10-deacetylbaccatin thiosemicarbazone. Inorganic Chemistry Communications. 1999;**2**:545-548

[54] Miller MC, Stineman CN, Vance JR, West DX, Hall IH. Multiple mechanisms for cytotoxicity induced by copper(II) complexes of 2-acetylpyrazine-N-substituted thiosemicarbazones. Applied Organometallic Chemistry. 1999;**13**:9-19

[55] Belicchi Ferrari M, Fava GG, Leporati E, Pelosi G, Rossi R, Tarasconi P, et al. Synthesis, characterization and biological activity of three copper(II) complexes with a modified nitrogenous base: 5-formyluracil thiosemicarbazone.

Journal of Inorganic Biochemistry. 1998;**70**:145-154

[56] Rogolino D, Cavvvazzoni A, Gatti A, Tegoni M, Pelosi G, Verdolino V, et al. Anti-proliferative effects of copper(II) complexes with hydroxyquinoline-thiosemicarbazone ligands. European Journal of Medicinal Chemistry. 2017;**128**:140-153

[57] Belicchi Ferrari M, Bisceglie F, Pelosi G, Tarasconi P, Albertini R, Pinelli S. New methyl pyruvate thiosemicarbazones and their copper and zinc complexes: Synthesis, characterization, X-ray structures and biological activity. Journal of Inorganic Biochemistry. 2001;**87**:137-147

[58] Muralisankar M, Haribabu J, Bhuvanesh NSP, Karvembu R, Sreekanth A. Synthesis, X-ray crystal structure, DNA/protein binding, DNA cleavage and cytotoxic studies of N(4) substituted thiosemicarbazone based copper(II)/nickel(II) complexes. Inorganica Chimica Acta. 2016;**449**:82-95

[59] Jagadeesh M, Lavanya M, Kalangi SK, Sarala Y, Ramachandraiah C, Varada Reddy A. Spectroscopic characterization, antioxidant and antitumor studies of novel bromo substituted thiosemicarbazone and its copper(II), nickel(II), and palladium(II) complexes. Spectrochimica Acta Part A. 2015;**135**:180-184

[60] Bisceglie F, Pinelli S, Alinovi R, Goldoni M, Mutti A, Camerini A, et al. Cinnamaldehyde and cuminaldehyde thiosemicarbazones and their copper (II) and nickel(II) complexes: A study to understand their biological activity. Journal of Inorganic Biochemistry. 2014;**140**:111-125

[61] Jagadeesh M, Kalangi SK, Krishna LS, Varada Reddy A. Halo-substituted thiosemicarbazones and their copper(II), nickel(II) complexes: Detailed spectroscopic characterization and study of antitumor activity against HepG2 human hepatoblastoma cells. Spectrochimica Acta Part A. 2014;**118**:552-556

[62] Shao J, Ma ZY, Li A, Liu YH, Xie CZ, Qiang ZY, et al. Thiosemicarbazone Cu(II) and Zn(II) complexes as potential anticancer agents: Syntheses, crystal structure, DNA cleavage, cytotoxicity and apoptosis induction activity. Journal of Inorganic Biochemistry. 2014;**136**:13-26

[63] Pelosi G, Bisceglie F, Bignami F, Ronzi P, Schiavone P, Carla Re M, et al. Antiretroviral activity of thiosemicarbazone metal complexes. Journal of Medicinal Chemistry. 2010;**53**:8765-8769

[64] Md Yusof EN, Ravoof TBSA, Tiekink ERT, Veerrakumarasivam A, Crouse KA, Mohamed Tahir MI, et al. Synthesis, characterization and biological evaluation of transition metal complexes derived from N,S bidentate ligands. International Journal of Molecular Sciences. 2015;**16**:11034-11054

[65] Selvamurugan S, Ramachandran R, Vijayan P, Manikandan R, Prakash G, Viswanatha murthi, P., Velmururgan, K., Nandhakumar, R., Endo, A. Synthesis, crystal structure and biological evaluation of Ni(II) complexes containing 4-chromone-*N*(4)-substituted thiosemicarbazone ligands. Polyhedron. 2016;**107**:57-67

[66] Palanimuthu D, Samuelson AG. Dinuclear zinc bis(thiosemicarbazone) complexes: Synthesis, in vitro anticancer activity, cellular uptake and DNA interaction study. Inorganica Chimica Acta. 2013;**408**:152-161

[67] Prabhakaran R, Kalaivani P, Huang R, Poornima P, Padma VV, Dallemer F, et al. DNA binding, antioxidant, cytotoxicity (MTT, lactate dehydrogenase, NO), and cellular uptake studies of structurally different nickel(II) thiosemicarbazone complexes: Synthesis, spectroscopy, electrochemistry, and X-ray crystallography. Journal of Biological Inorganic Chemistry. 2013;**18**:233-247

[68] Afrasiabi Z, Sinn E, Lin W, Ma Y, Campana C, Padhye S. Nickel(II) complexes of naphthaquinone thiosemicarbazone and semicarbazone: Synthesis, structure, spectroscopy, and biological activity. Journal of Inorganic Biochemistry. 2005;**99**:1526-1531

[69] Belicchi Ferrari M, Capacchi S, Reffo G, Pelosi G, Tarasconi P, Albertini R, et al. Synthesis, structural characterization and biological activity of *p*-fluoro benzaldehyde thiosemicarbazones and of a nickel complex. Journal of Inorganic Biochemistry. 2000;**81**:89-97

[70] Gou Y, Wang J, Chen S, Zhang Z, Zhang Y, Zhang W, et al. α-N-heterocyclic thiosemicarbazone Fe(III) complex: Characterization of its antitumor activity and identification of anticancer mechanism. European Journal of Medicinal Chemistry. 2016;**123**:354-364

[71] Richardson DS, Kallinowshi DS, Richardson V, Sharpe PC, Lovejoy DB, Islam M, et al. 2-Acetylpyridine thiosemicarbazones are potent iron chelators and antiproliferative agents: Redox activity, iron complexation and characterization of their antitumor activity. Journal of Medicinal Chemistry. 2009;**52**:1459-1470

[72] Kowol CR, Berger R, Eichinger R, Roller A, Jakupec MA, Schmidt PP, et al. Gallium(III) and iron(III) complexes of α-N-heterocyclic thiosemicarbazones: Synthesis, characterization, cytotoxicity, and interaction with ribonucleotide reductase. Journal of Medicinal Chemistry. 2007;**50**:1254-1265

[73] Richardson DR, Sharpe PC, Lovejoy DB, Senaratne D, Kalinowski DS, Islam M, et al. Dipyridyl thiosemicarbazone chelators with potent and selective antitumor activity form iron complexes with redox activity. Journal of Medicinal Chemistry. 2006;**49**:6510-6521

[74] Mohan M, Kumar M, Kumar A, Madhuranath PH, Jha NK. Synthesis, characterization, and antitumour activity of iron(II) and iron(III) complexes of α-heterocyclic carboxaldehyde thiosemicarbazones. Journal of Inorganic Biochemistry. 1988;**32**:239-249

[75] Salam MA, Hussein MA, Ramli I, Islam MS. Synthesis, structural characterization and evaluation of biological activity of organotin(IV) complexes with 2-hydroxy-5-methoxy benzaldehyde-N(4)-methylthiosemicarbazone. Journal of Organometallic Chemistry. 2016;**813**:71-77

[76] Li MX, Zhang D, Zhang LZ, Niu JY, Ji BS. Diorganotin(IV) complexes with 2-benzoylpyridine and 2-acetylpyrazine N(4)-phenylthiosemicarbazones: Synthesis, crystal structures and biological activities. Journal of Organometallic Chemistry. 2011;**696**:852-858

[77] Casas JS, Rodriguez-Arguelles MC, Russo U, Sanchez A, Sordo J, Vazquez-Lopez A, et al. Diorganotin(IV) complexes of pyridoxal thiosemicarbazone: Synthesis, spectroscopic properties and biological activity. Journal of Inorganic Biochemistry. 1998;**69**:283-292

[78] Tai YX, Ji YM, Lu YL, Li MX, Wu YY, Han QX. Cadmium(II) and

indium(IIII) complexes derived from 2-benzoylpyridine *N*(4)-cyclohexylthiosemicarbazone: Synthesis, crystal structures, spectroscopic characterization and cytotoxicity. Synthetic Metals. 2016;**219**:109-114

[79] Vijayan P, Viswanathamurthi P, Silambarasan V, Velmurugan D, Velmurugan K, Nandhakumar R, et al. Dissymmetric thiosemicarbazone ligands containing substituted aldehyde arm and their ruthenium(II) carbonyl complexes with PPh3/AsPh3 as ancillary ligands: Synthesis, structural characterization, DNA/BSA interaction and in vitro anticancer activity. Journal of Organometallic Chemistry. 2014;**768**:163-177

[80] Kumar K, Schniper S, Gonzalez-Sarrias A, Holder AA, Sanders N, Sullivan D, et al. Highly potent anti-proliferative effects of a gallium(III) complex with 7-chloroquinoline thiosemicarbazone as a ligand: Synthesis, cytotoxic and antimalarial evaluation. European Journal of Medicinal Chemistry. 2014;**86**:81-86

[81] Parrilha GL, Ferraz KSO, Lessa JA, Navakoski de Oliveira K, Rodrigues BL, Ramos JP, et al. Metal complexes with 2-acetylpyridine-N(4)-orthochlorophenylthiosemicarbazone: Cytotoxicity and effect on the enzymatic activity of thioredoxin reductase and glutathione reductase. European Journal of Medicinal Chemistry. 2014;**84**:537-544

[82] Yousef TA, Abu El-Reash GM, Al-Jahdali M, El-Rakhawy EBR. Synthesis, spectral characterization and biological evaluation of Mn(II), Co(II), Ni(II), Cu(II), Zn(II) and Cd(II) complexes with thiosemicarbazone ending by pyrazole and pyridyl rings. Spectrochimica Acta Part A. 2014;**129**:163-172

[83] Manikandan R, Chitrapriya N, Jang YJ, Viswanathamurthi P. Evaluation of DNA-binding, radical scavenging and cytotoxic activity of five coordinated Cd(II) complexes containing 2-acetylpyridine-N4-substituted thiosemicarbazone. RSC Advances. 2013;**3**:11647-11657

[84] Li MX, Yang M, Niu JY, Zhang LZ, Xie SQ. A nine-coordinated bismuth(III) complex derived from pentadentate 2,6-diacetyl pyridine bis(^{4}N-methylthiosemicarbazone): Crystal structure and both in vitro and in vivo biological evaluation. Inorganic Chemistry. 2012;**51**:12521-12526

[85] Stringer T, Therrien B, Hendricks DT, Guzgay H, Smith GS. Mono- and dinuclear (η6-arene) ruthenium(II) benzaldehyde thiosemicarbazone complexes: Synthesis, characterization and cytotoxicity. Inorganic Chemistry Communications. 2011;**14**:956-960

[86] Li MX, Chen CL, Zhang D, Niu JY, Ji BS. Mn(II), Co(II) and Zn(II) complexes with heterocyclic substituted thiosemicarbazones: Synthesis, characterization, X-ray crystal structures and antitumor comparison. European Journal of Medicinal Chemistry. 2010;**45**:3169-3177

[87] Kovala-Demertzi D, Papageorgiou A, Papathanasis L, Alexandratos A, Dalezis P, Miller JR, et al. In vitro and in vivo antitumor activity of platinum(II) complexes with thiosemicarbazones derived from 2-formyl and 2-acetyl pyridine and containing ring incorporated at N(4)-position: Synthesis, spectroscopic study and crystal structure of platinum(II) complexes with thiosemicarbazones, potential anticancer agents. European Journal of Medicinal Chemistry. 2009;**44**:1296-1302

[88] Noblia P, Vieites M, Parajon-Costa BS, Baran EJ, Cerecetto H, Draper P, et al. Vanadium(V) complexes

with salicylaldehyde semicarbazone derivatives bearing in vitro anti-tumor activity toward kidney tumor cells (TK-10): Crystal structure of [V^VO_2(5-bromosalicylaldehde semicarbazone)]. Journal of Inorganic Biochemistry. 2005;**99**:443-451

[89] Afrasiabi Z, Sinn E, Chen J, Ma Y, Rheingold AL, Zakharov LN, et al. Appended 1,2-naphthoquinones as anticancer agents 1: Synthesis, structural, spectral and antitumor activities of ortho-naphthaquinone thiosemicarbazone and its transition metal complexes. Inorganica Chimica Acta. 2004;**357**:271-278

[90] Arion VB, Jakupec MA, Galanski M, Unfried P, Keppler BK. Synthesis, structure, spectroscopic and in vitro antitumor studies of a novel gallium(III) complex with 2-acetyl pyridine ^{4}N-dimethylthiosemicarabzone. Journal of Inorganic Biochemistry. 2002;**91**:298-305

[91] Perez JM, Matesanz AI, Ambite AM, Navarro P, Alonso C, Souza P. Synthesis and characterization of complexes of *p*-isopropyl benzaldehyde and methyl 2-pyridyl ketone thiosemicarbazones with Zn(II) and Cd(II) metallic centers. Cytotoxic activity and induction of apoptosis in Pam-ras cells. Journal of Inorganic Biochemistry. 1999;**75**:255-261

[92] Jayasree S, Aravindakshan KK. Synthesis, characterization and antitumour studies of metal chelates of acetoacetanilide thiosemicarbazone. Transition Metal Chemistry. 1993;**18**:85-88

[93] Mohan M, Agarawal A, Jha NK. Synthesis, characterization and antitumour properties of some metal complexes of 2,6-diacetylpyridine bis(N^4-azacyclic thiosemicarbazones). Journal of Inorganic Biochemistry. 1988;**34**:41-54

Chapter 8

Recent Advancements in Schiff Base Fluorescence Chemosensors for the Detection of Heavy Metal Ions

Sabeel M. Basheer, Puthiyavalappil Rasin, Vipin Manakkadan, Vishnunarayanan Namboothiri Vadakkedathu Palakkeezhillam and Anandaram Sreekanth

Abstract

The Schiff base was first synthesized by Hugo Schiff through the condensation reaction of primary amines with carbonyl compounds (aldehyde or ketone) in 1864. Schiff bases exhibit many structural and electrical characteristics that enable their use in a variety of fields, including medical and chemosensing. Schiff bases generate stable complexes when they bind with different metal ions. Schiff bases are employed as fluorescent turn-on/turn-off chemosensors for the detection of various metal cations, such as Hg^{2+}, Cd^{2+}, Cr^{3+}, Pd^{2+}, and As^{3+} in various materials due to their outstanding coordination ability. This chapter examines a variety of Schiff bases that are employed in chemosensing procedures for various metal ions (such as divalent and trivalent cations) in various biological, agricultural, and environmental settings.

Keywords: thiosemicarbazones, chemosensor, fluorescence, copper, fluoride ion

1. Introduction

Various fields use metallic cations, some of which control countless biological processes necessary for life. However, since they are not biodegradable and can accumulate in the food chain, the excess of these ions can lead to serious environmental issues. Even at low concentrations, they constitute a substantial threat to both the environment and human health. Numerous health issues, such as allergies, lung damage, anemia, kidney failure, neurotoxicity, genotoxicity, oxidative toxicity, steroidogenic toxicity, sperm toxicity, apoptotic toxicity, and axillary toxicity, can be brought on by these ions [1–10]. So it is crucial to architect an effective method for identifying these ions in various samples. In order to detect metallic cations, different techniques have been developed, including liquid chromatography [11], electrochemistry [12], voltammetry [13], reversed-phase high-performance liquid chromatography [14], and inductively coupled plasma-mass spectrometry [15]. Even though these methods are

 IntechOpen

quite effective, they nevertheless have certain drawbacks, such as being expensive, difficult to use, requiring a lot of sample preparation, and providing ambient conditions for their use. As an alternative to the aforementioned drawbacks of conventional techniques, fluorescent-based Schiff bases have been developed for the detection of metallic cations. They serve as stabilizers for polymers, catalysts, pigments, and dyes, intermediates in organic synthesis, and pigments and dyes [16]. Additionally, a variety of biological activities involving antifungal, antibacterial, antimalarial, antiproliferative, anti-inflammatory, antiviral, and antipyretic effects have been linked to Schiff bases [17, 18]. There are imine or azomethine groups in a variety of natural, natural-derived, and synthetic substances. It has been demonstrated that these compounds' imine group is essential for their biological actions. Schiff bases with great sensitivity and selectivity to a variety of species (cations and anions) were demonstrated by the spectrofluorometric method. In particular, chemosensors based on Schiff bases have shown exceptional performance for the identification of different metallic cations due to their simple and affordable production as well as their capacity to coordinate with practically all metal ions and stable them in a variety of oxidation states. Additionally, these Schiff bases have demonstrated a wide range of biological uses, such as anti-ulcerogenic, analgesic, antifungal, anti-inflammatory, anti-viral, antioxidant, and anticancer effects. In various processes, many of these compounds exhibit outstanding catalytic efficiency [19, 20]. Various Schiff bases have been examined in this review article as fluorescent turn-on/turn-off chemosensors for the detection of different metal ions in various matrices. Additionally, many Schiff base chemosensors' synthetic processes and sensing mechanisms have been studied.

2. Synthesis of Schiff bases

Any primary amine can produce Schiff bases under specified conditions when it interacts with an aldehyde or ketone. In terms of structure, a Schiff base is an aldehyde or ketone that has had the carbonyl group (C=O) substituted by an imine or azomethine group, creating a nitrogen analogue. The synthesis of Schiff bases has used a variety of synthetic methods. Only a few straightforward synthetic methods are described here for the benefit of beginners and others who are unfamiliar with organic synthetic chemistry. Hugo Schiff (1864) was the first to produce Schiff bases using an azeotropic distillation reaction between primary amines and an aldehyde or ketone (Compound a, **Figure 1**) [21]. Catalysts of many types, such as acetic acid, p-toluene sulfonic acids, montmorillonite, and acid resin. There have been reports of the synthesis of Schiff bases from primary amines and carbonyl compounds (aldehyde or ketone) using resin, HCl, H_2SO_4, $TiCl_4$, $Mg(ClO_4)_2$, $SnCl_4$, BF_3Et_2O, $ZnCl_2$, $MgSO_4$, etc. Additionally, under green chemistry circumstances, microwave irradiation of Schiff bases has been modified for solvent-free synthesis [22, 23]. In this approach, aldehyde or ketone-containing carbonyl compounds react with primary amines to produce Schiff bases with a high yield (Compound b, **Figure 1**). The process can proceed at ambient by the use of an appropriate catalyst. Without the use of any solvent, the reactants (primary amines and aldehyde/ketone) are ground in a mortar and pestle (Compound c, **Figure 1**) [24]. Additionally, primary amines and alcohol can be combined to create a Schiff base (Compound d, **Figure 1**). This approach works with a variety of alcohols and amines and does not need any additional equipment or supplements [25].

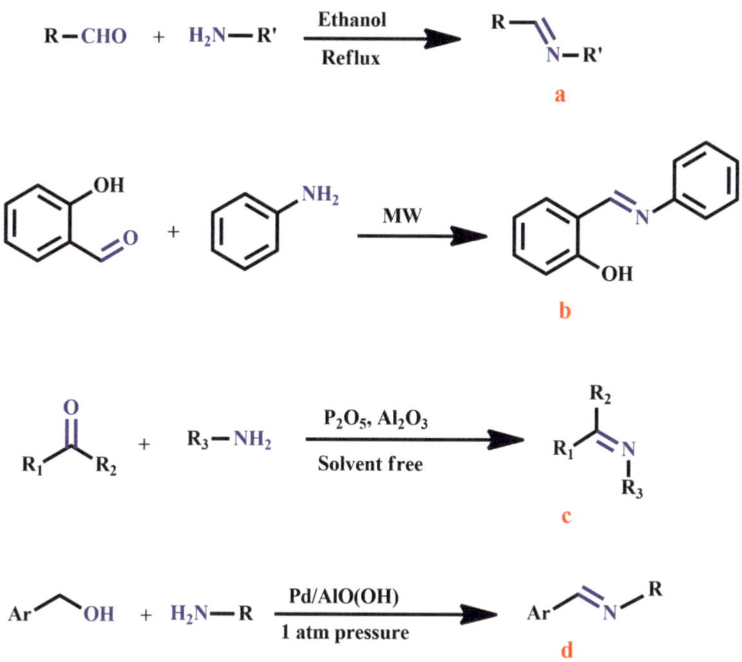

Figure 1.
The synthesis of Schiff base derivatives via several synthetic pathways.

3. Chemosensors using Schiff bases to find hazardous metal ions

For fluorescence turn-on/turn-off detection of different metal cations, a number of chemosensors based on Schiff bases have been created and employed (**Table 1**). Before employing these compounds as chemosensors, there are a few ideas to keep in mind. In order to create a heterogeneous system in an aqueous medium, bulky chains are preferred. This makes it possible to set up a system for quick and affordable separation. Additionally, spectroscopic techniques are used in the field of chemosensing; hence, the Schiff bases should preferably comprise a fluorophore for fluorescent spectroscopic studies [26]. Similar to this, a spacer (which can change the geometry of the system and control the electrical interaction between the receptor and chromophore unit) and a receptor (which is responsible for the selective analyte binding) are needed [27].

4. The sensing mechanism of the Schiff base chemosensor, including its selectivity and sensitivity

The electron configuration of both the ligand and the metal ion, as well as the structural stiffness and the capacity of the Schiff base chemosensor to attach to the metal ion, is a parameter that controls the selectivity and high sensitivity of Schiff base chemosensors [28, 29]. Therefore, various mechanisms, such as chelation-enhanced fluorescence (CHEF), photo-induced electron transfer (PET), intra/intermolecular charge transfer (ICT), hydrolysis, chelation enhancement quenching effect

Sensor	Sensor type	Sensing mechanism	Analyte	M: L	LOD	Wavelength $\lambda ex/\lambda em$ (nm)	pH range	Matrices	Ref.
1	Turn on	Chelation & CHEF	Hg^{2+}	2:1	2.82 µM	347/445	6–8	Living cells	38
2	Turn on	—	Hg^{2+}	1:2	75 µM	369/464	8	Ethanol/H_2O	39
3	Turn off	PET	Hg^{2+}	1:2	8.3 nM	420/478	3	HEPES buffer	40
4	Turn on	—	Hg^{2+}	1:1	55.95 nM	305/440	—	DMSO	41
5	Turn off	ICT	Hg^{2+}	1:1	25.2 nM	370/436	—	can	42
6	Turn on	ESIPT	Hg^{2+}	1:2	22.6 nM	349/473	8	H_2O/AcN	43
7	Turn on	CHEF	Cd^{2+}	1:2	14.8 nM	456/510	—	H_2O/AcN	44
8	Turn on	PET/CHEF	Cd^{2+}	1:1	0.12 µM	380/506	5-9	can	45
9	Turn on	FRET	Cd^{2+}	1:1	10.25 nM	308/590	3-9	can	46
10	Turn on	—	Cd^{2+}	1:1	60 µM	385/474	8	Ethanol/H_2O	47
11	Turn on	CIFA	Cd^{2+}	1:1	0.218 µM	530/591	—	H_2O	48
12	Turn on	PET	Cr^{3+}	1:1	1.5 µM	320/562	8	H_2O/THF	49
13	Turn on	Dissociation	Cr^{3+}	1:2	0.94 nM	327/418	5-10	Ethanol	50
14	Turn off	IFE	Cr^{6+}	1:1	0.175 µM	350/449	—	Ascorbicacid/ H_2O	51
15	Turn on	PET	Cr^{3+}	1:2	0.5 µM	279/568	6-10	MeOH/H_2O	52
16	Turn on	PET	Cr^{3+}	1:1	13 µM	360/663	9	H_2O/AcN	53
17	Turn on	Chelation	Pd^{2+}	1:1	0.05 µM	485/563	7.4	living cells HCT116	54
18	Turn on	Chelation	Pd^{2+}	1:1	0.018 µM	450/598	7–12	living cells MCF7	55
19	Turn on	ESIPT	As^{3+}	1:1	69.7 nM	410/525	1-13	Naphtha/Ethanol	59
20	Turn on	—	As^{3+}	1:1	7.2 ppb	350/–	6-8	H_2O	60
21	Turn on	—	As^{3+}	1:1	0.45 µM	385/496	—	THF	61
22	Turn on	Chemodosimetry	Au^{3+}	1:1	1.51 µM	560/588	—	H_2O/AcN	62
23	Turn on	AIE	Y^{3+}	1:1	22.3 µM	360/575	4-8	THF/H_2O	63

Sensor	Sensor type	Sensing mechanism	Analyte	M: L	LOD	Wavelength λex/λem (nm)	pH range	Matrices	Ref.
24	Turn on	ICT	Y^{3+}	1:1	0.013 ppb	435/495	—	THF	64
25	Turn on	PET	Ga^{3+}	1:1	0.01 μM	300/434	—	DMF/AcN	65
26	Turn on	PET	Ga^{3+}	1:2	3.90 nM	352/421	—	H_2O/AcN	66
27	Turn on	PET	Ga^{3+}	1:2	3.97 nM	352/421	—	H_2O/AcN	67

Table 1.
Different analytical parameters for the analysis of heavy metal ions using Schiff bases.

(CHEQ), and ring-opening mechanisms, have been proposed for the interaction between metal ions and various chemosensors. In this chapter, we also briefly address the mechanisms underlying the chemosensory applications of Schiff base derivatives. (i) Electron transfer (ET): Out of all of these traditional processes, ET-primarily via photo-induced electron transfer (PET)—has been the one used the most frequently in fluorescent chemosensors. A fluorophore's fluorescence is typically quenched by the PET process; however, it may be restored if guest molecules can prevent the PET process. (ii) Charge transfer (CT): ICT (intramolecular charge transfer), MLCT (metal-ligand charge transfer), and twisted ICT are examples of CT processes (TICT). A ratiometric signal is produced by an ICT-based chemosensor when an ICT process is enhanced or suppressed. This ratiometric signal can remove the majority of ambiguities by self-calibrating two emission bands and enable quantitative determination in more advanced applications, such as imaging in living cells and tissues. On the other hand, transition metal complexes like those of ruthenium, rhenium, and iridium, etc., frequently exhibit MLCT, in which CT occurs from a ligand to a transition metal cation. Through the influence of MLCT energy level by analytes, it can also be employed for chemosensor design. Additionally, TICT is a potent intramolecular CT that takes place in the excited state and requires solvent relaxation surrounding the molecule to produce an ongoing rotation of the electron donor and acceptor until it is twisted roughly 90 degrees [30, 31]. Since intramolecular rotation and charge separation in the TICT state depend on polar solvent relaxation, the fluorescence behavior is extremely sensitive to micropolarity and/or steric barrier for molecular rotation [29, 32, 33]. (iii) Energy transfer (ET): Depending on the interaction distance between the energy donor and energy acceptor, ET can be divided into two categories: electronic energy transfer (EET) and fluorescence resonance energy transfer (FRET). While FRET requires a specific amount of spectral overlap between the emission spectrum of the donor and the acceptor, EET, also known as Dexter electron transfer, requires a distance between the donor and acceptor of less than 10 to be effective. For effective FRET to happen, the distance between the donor and acceptor needs to be between 10 and 100. As a result, the chemosensors based on the energy transfer mechanism are highly dependent on distance. (iv) Excimer/exciplex: An excimer is a complex that is created when a fluorophore interacts with another fluorophore of the same structure in the ground state while it is stimulated. The resulting complex is known as an exciplex if the fluorophore in the excited state differs from the fluorophore in the ground state. A dual emission from the monomer and excimer/exciplex is frequently recorded at the same time because the emission spectra of an excimer/exciplex are red-shifted in comparison to that of the monomer. The excimer/exciplex band can therefore be monitored to detect excimer/exciplex production or deformation in response to interaction with a guest species. However, there are currently just a few examples of chemosensors that work by forming exciplexes [32, 34, 35].

5. Detection of some toxic heavy metal ions

5.1 Detection of mercury

Mercury is a toxic heavy metal even at low concentrations and widely dispersed in nature. It can accumulate in the body and is responsible for different diseases including acute kidney failure, prenatal brain damage, and heart diseases. Therefore, the detection of Hg^{2+} ions in different samples is of great consideration. A novel Schiff base chemosensor 1 containing pyrene-based free thiol derivative was used for the determination of Hg^{2+} ions in living cells [36]. Therefore, it is quite important to consider finding Hg^{2+} ions in various samples. For the detection of Hg^{2+} ions in live cells, a new Schiff base chemosensor 1 with pyrene-based free thiol derivative was employed. Similar to this, a double-naphthalene Schiff base fluorescent chemosensor 2 was used to detect Hg^{2+} ions at a low detection limit of 0.05595 μM in a DMSO medium (**Figure 2**). In the presence of DMSO, the chemosensor's imine group underwent oxidation to become an amide. The 1:1 metal-ligand complex generated by the N- and O- atoms of amide and Hg^{2+} ions was confirmed by mass spectrometry, ^{1}H NMR, and FT-IR spectroscopy [37]. A novel coumarin-based fluorescent chemosensor 3 was employed in another study to find Hg^{2+} ions. In the presence of secondary metal ions, the chemosensor shows excellent selectivity toward Hg^{2+}. As the amount of Hg^{2+} gradually increases, the probe's yellow-colored solution becomes colorless. Fluorescence investigations further supported chemosensor 3's selectivity for mercury ions. The bright blue fluorescence at 460 nm gradually decreases in conjunction with a minor red-shift to the cyan channel (470 nm) after the addition of 0.4 M Hg^{2+} to the probe solution. The lone pair electrons are distributed to Hg^{2+} by the Schiff base imine nitrogen (CH=N) and phenolic-oxygen atoms, which reduces the emission of acceptor coumarin and prolongs that of the donor 4-(Diethylamino)salicylaldehyde. Therefore, 4-(Diethylamino)salicylaldehyde's distinctive emission at 470 nm is due to the inhibition of the ESIPT and PET-ON processes. The mole fraction at maximum absorbance is 0.3, showing that Cou-S and both Hg^{2+} form a dimer complex with a binding stoichiometry of 2:1 (probe: metal ion). An 8.3 nM threshold for detection was determined. Density functional theory (DFT) was also used to analyze how the chemosensor and Hg^{2+} ions interact [38]. Another work used a novel N-salicylidene) benzylamine Schiff base 4 to detect Hg^{2+} ions via fluorescence ON characteristics. Fluorescence and UV-Vis spectroscopy were used to examine the chemosensor's sensing capabilities. By using fluorescence emission and UV-Vis spectroscopy, it was determined how the chemosensor 4 responded to different cations. Significant fluorescence enhancement and spectral/color alterations supported the sensor's selectivity for Hg^{2+} [39]. The detection of Hg^{2+} ions is influenced by the pH of the detection medium. For example, a turn-on fluorescent chemosensor 5 was developed to detect Hg^{2+} in an aqueous CH_3CN medium with a pH range of 5–10 [40]. In the presence of various cations including Hg^{2+}, Pb^{2+}, Ca^{2+}, Ba^{2+}, Cu^{2+}, Cr^{3+}, Mg^{2+}, Zn^{2+} ions in aqueous media, it was discovered that the Hg^{2+} ion notably exhibited specificity for the Schiff base chemosensor 6. Additionally, counter anions such as Br^-, Cl^-, F^-, I^-, S^{2-}, CN^-, NO_3^-, NO_2^-, OH^-, CO_3^-, SCN^-, HSO_3^-, $H_2PO_4^-$ CH_3COO^-, HPO_4^{2-}, $P_2O_7^{4-}$, and PO_4^{3-} failed to have a discernible impact on the chemosensor's selectivity [41]. The structures of the chemosensors 1–6 and their interaction with Hg^{2+} ion are shown in **Figure 2**.

5.2 Detection of cadmium ion

Due to the extensive range of uses of cadmium in numerous industries, including mining, smelting, and the combustion of fossil fuels, etc., it is one of the most pervasive health-hazardous contaminants. In light of this, it is crucial to create a reliable and efficient method for detecting Cd^{2+} ions in a variety of materials. The quinoline moiety-containing Schiff base chemosensor 7 was created as a fluorescence sensor for Cd^{2+} ions. It's interesting to note that the chemosensor performed superbly when detecting Cd^{2+} ions in an aqueous media by turn-on fluorescence.

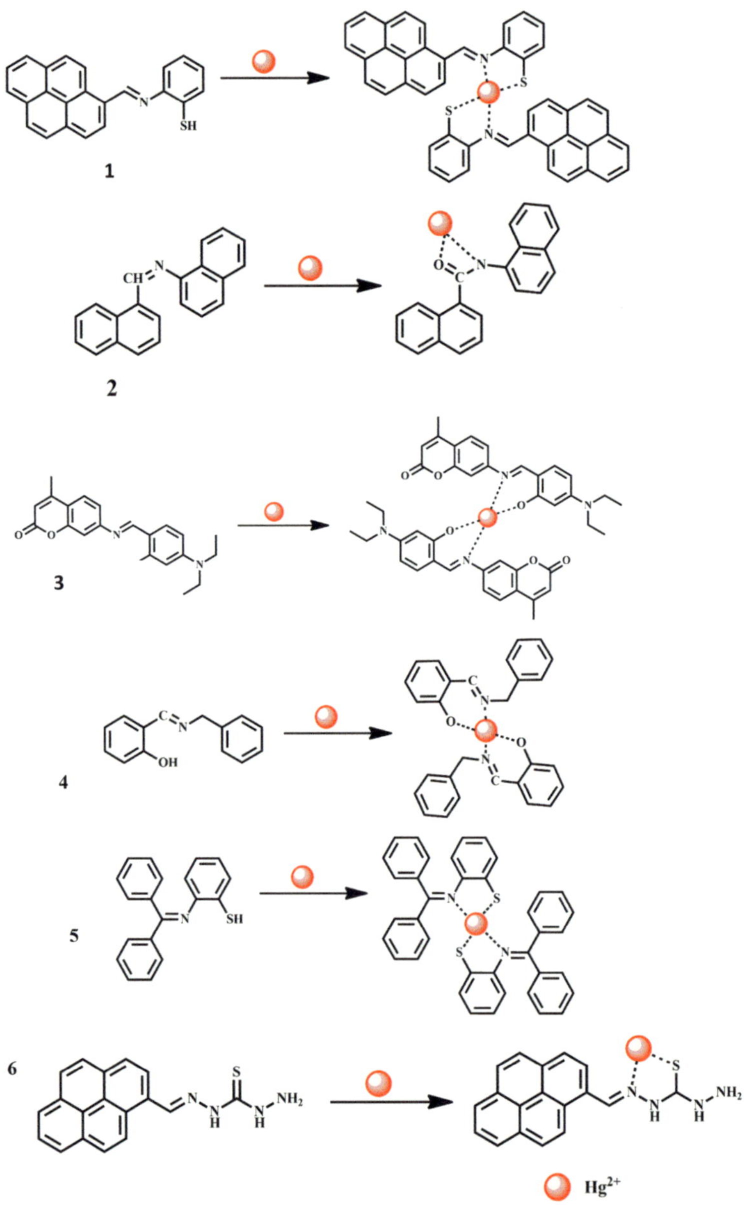

Figure 2.
Formation of Hg^{2+} complexes with Schiff base chemosensors 1–6.

Very little fluorescence emission was seen from chemosensor 7 when it was unmodified, but a significant turn-on response was seen when Cd^{2+} was added. The sensor showed excellent selectivity with no interference from any other metal ions. The limit of detection was determined to be 0.0024 μM, and the binding stoichiometry between the Cd^{2+} and chemosensor 7 was proven to be 1:1 [42]. To identify Cd^{2+} ions, a pyridine-based Schiff base chemosensor 8 was created. The fluorescence enhancement happened when the Cd^{2+} ion was applied to the complexion despite the chemosensor not being fluorescent. As a result, the CHEF was stimulated while the PET was blocked, increasing the fluorescence intensity of the chemosensor. Additionally, chemosensor 8 was successfully used in water sample analysis [43]. With a detection limit of 1.025×10^{-8} M, a Rhodamine-based Schiff-base fluorescence chemosensor 9 was used to find Cd^{2+} ions. In the presence of cations, the chemosensor 9 shows significant sensitivity and selectivity for Cd^{2+} ions [44]. An ON-type Schiff base chemosensor 10 was communicated in 2020 and its fluorescence properties against cadmium ions were discussed. According to Job plot, the chemosensor creates a complex with Cd^{2+} of 1:1 ratio. For an analytical approach based on chemosensor, pH, solvent type, and ligand concentration were tuned for the detection of Cd^{2+} in aqueous samples. The limit of detection was determined to be 6.0×10^{-7} M. The probe showed a wide range of linearity with Cd^{2+} [45]. A recent study described fluorescent sensor 11, which is based on rhodamine, and its capacity to find Cd^{2+} ions. The coordination-induced fluorescence activation (CIFA) mechanism is how the sensor reacts to Cd^{2+}. In the presence of the tested metal ions, chemosensor 11 responds to Cd^{2+} with a highly quick and reversible fluorescence. The complex stoichiometry between the sensor and Cd^{2+} was discovered to be 1:1, and the binding constant in acetonitrile (ACN)/HEPES buffer (10 mM, pH, 7.05, v/v 1:1) was calculated to be 2.70 10^7 M^{-1}. The chemosensor 11's fluorescence detection limit for Cd^{2+} was discovered to be 0.218 μM, demonstrating a notable sensitivity to Cd^{2+} [40]. The structures of the chemosensors 7–11 and their interaction with Cd^{2+} ion are shown in **Figure 3**.

5.3 Detection of chromium ion

Chromium is a significant transition metal that is used in a variety of industries, including chemical engineering, textile manufacturing, steel production, oil refining, and electroplating. But it raises the risk of lung, sinus, and nasal cancer and results in pulmonary sensitization. A novel Schiff base chemosensor 12 was constructed and subjected to several characterization methods in order to detect Cr^{3+}. Using a spectrophotometric method, the chemosensor was tested against a variety of toxic metal ions, including Mn^{2+}, Ag^{+}, Cd^{2+}, Co^{2+}, Cu^{2+}, Ni^{2+}, Zn^{2+}, Fe^{3+}, Al^{3+}, and Cr^{3+}. Only Cr^{3+} ions were found to produce a recognizable hyperchromic shift in the chemosensor's absorbance due to the formation of the stable complex through S and N atom [46].

Additionally, electrochemical methods were employed to investigate the electrochemical properties and reversibility of the chemosensor. To detect Cr^{3+} ions in the aqueous media, the Schiff base chemosensor 13 was also used. Visual monitoring was done of the chemosensor 13's colorimetric sensitivity to several cations. When chromium ions were added, the chemosensor 13's yellow color turned to colorless. Fluorescence tests were used to investigate the chemosensor 13's sensing capabilities in more detail. The chemosensor 13 displayed less fluorescence intensity at 418 nm, but a significant increase in fluorescence intensity was seen in the presence of Cr^{3+} due to the imine group of sensor might be converted back to the carbonyl group in the presence of Cr^{3+} [47]. Recently, a novel Schiff base fluorescent chemosensor 14 for the selective detection of chromium ions was published. The development of a 1:1 stoichiometric sensor-Cr^{3+} complex, which is supported by Job's plot, has the effect of inhibiting the PET process. The limit of detection was shown to be 1.3×10^{-7} M via fluorescence titration, and the association constant Ka was reported as 2.28×10^{5} M^{-1}. Additionally, DFT and TDDFT simulation results were provided to understand the proposed complex's optimal structure as well as its electronic spectra [48]. An innovative thiazole-based fluorescent Schiff base chemosensor 15 for the chemodosimetric method to Cr(III) ion detection was discussed. The structure of the chemosensor 15 was determined using a variety of analytical techniques, including UV-vis, ^{13}C-NMR,

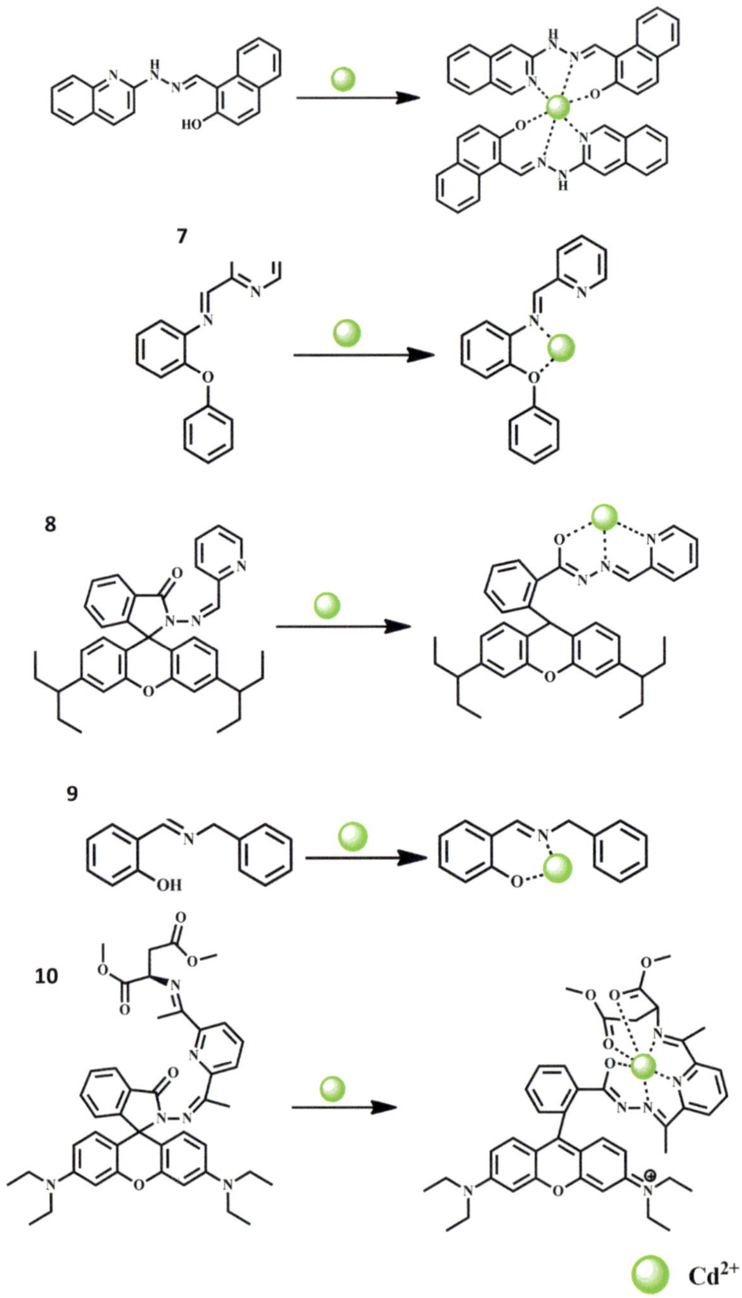

Figure 3.
Formation of Cd^{2+} complexes with Schiff base chemosensors 7–11.

^{1}H-NMR, and FT-IR analysis. It's interesting to note that chemosensor 15 responds to different metal cations, such as Ni^{2+}, Na^{+}, Cd^{2+}, Ag^{+}, Mn^{2+}, K^{+}, Zn^{2+}, Cu^{2+}, Hg^{2+}, Co^{2+}, Pb^{2+}, Mg^{2+}, Sn^{2+}, Al^{3+}, and Cr^{3+}, by turning on fluorescently to specific Cr(III) ions [49]. Whereas an inner filter effect-based Schiff base chemosensor 16 was adequately quenched by Cr(VI) through primary and secondary inner filter effects. The addition of L-ascorbic acid in the concentration range of 10 μM to 390 μM with an LOD of 2.46 μM efficiently turned on the chemosensor-Cr(VI) solution's switched-off fluorescence. The reduction of Cr(VI) to Cr(III) by L-ascorbic acid led to the elimination of both primary and secondary inner filter effects and the recovery of the chemosensor's fluorescence, which is the mechanism proposed for the fluorescence turn-on of the Cr(VI)-chemosensor's quenched fluorescence [50]. The structures of the chemosensors 12-15 and their interaction with Cr^{3+} ion are shown in **Figure 4**.

5.4 Detection of palladium ion

A valuable metal, palladium is used in several industrial and electronic devices. However, palladium is extremely carcinogenic and poisonous. The detection of Pd^{2+} ions in various environmental, biological, and agricultural materials is therefore crucial. For the purpose of detecting Pd^{2+} ions, the Schiff base fluorescent and colorimetric chemosensor 17 was employed. High selectivity toward Pd^{2+} ions was demonstrated by this chemosensor in the presence of monovalent, divalent, trivalent, and tetravalent cations such as Ba^{2+}, Cu^{2+}, Zn^{2+}, Fe^{2+}, Tb^{3+}, Hg^{2+}, Eu^{3+}, Mg^{2+}, Gd^{3+}, Mn^{2+}, Sm^{3+}, Cr^{3+}, Ag^{+}, Ca^{2+}, Sn^{4+}, and Ni^{2+}. Mass, ^{1}H NMR, ^{13}C NMR, and FT-IR analyses all validated the spirolactam ring opening as the chemosensor's mechanism. Chemosensor 17's Pd^{2+} detection threshold was established at 0.05 μM [51]. In the presence of the ions Ba^{2+}, Cu^{2+}, Zn^{2+}, Al^{3+}, Fe^{2+}, Hg^{2+}, Mg^{2+}, Co^{3+}, Mn^{2+}, Ag^{+}, Ca^{2+}, Ni^{2+}, Cd^{2+}, Pb^{2+}, Fe^{3+}, Pt^{2+}, Na^{+}, and K^{+}, Coumarinyl-rhodamine Schiff base chemosensor 18 functions as an incredibly selective chemosensor for Pd^{2+}. The spirolactam ring being opened has been suggested as the mechanism for this chemosensor. The Pd^{2+} ions improve the fluorescence intensity and change the color of the chemosensor from straw color to pink, whereas the free chemosensor was weakly emissive. In the presence of chemosensor 18 and distinct metal cations, this chemosensor was highly selective for the Pd^{2+} ion [52]. The structures of the chemosensors 17–18 and their interaction with Cr^{3+} ion are shown in **Figure** 5.

5.5 Detection of Arsenic ion

Despite being extensively distributed in nature, arsenic is a deadly heavy metal even at low amounts. It builds up in the body and causes a variety of illnesses, such as lung and skin cancers, interference with cellular respiration, and affects the functions of the liver, kidney, bladder, and prostate [53–55]. Therefore, the detection of As3+ ions in different samples is of great consideration. A recent study described the detection of arsenic species found in a real naphtha sample using a fluorescence chemosensor 19 based on excited-state intramolecular proton transfer (ESIPT). The ESIPT process of M-HBT was demonstrated by the pH-related emission migration from 610 nm (pH 14,1) to 510 nm (pH 14,13). The sensing ability of M-HBT was examined in ethanol, and the experimental findings showed that As^{3+} responded linearly well in the range of 0–32 μM [56]. Similar to this, for the quick detection of As^{3+}/As^{5+} in mixed aqueous solutions, a novel colorimetric sensor based on a benzothiazole Schiff base chemosensor 20 has been reported. It offers advantages of simplicity, specificity, high sensitivity, selectivity, and economy with a detection limit of 7.0 ppb within 10 s via the naked eye. Moreover, the probe is appropriate for on-site, quick, and convenient detection of As^{3+}/As^{5+} ions at extremely low concentrations in actual water samples and exhibits remarkable selectivity in the presence of other common ions [57]. A novel coumarin-based fluorescent chemosensor 21 was employed in another study to find As^{3+} ions. In the presence of secondary metal ions (Na^{+}, Mg^{2+}, Cu^{2+}, Ni^{2+},

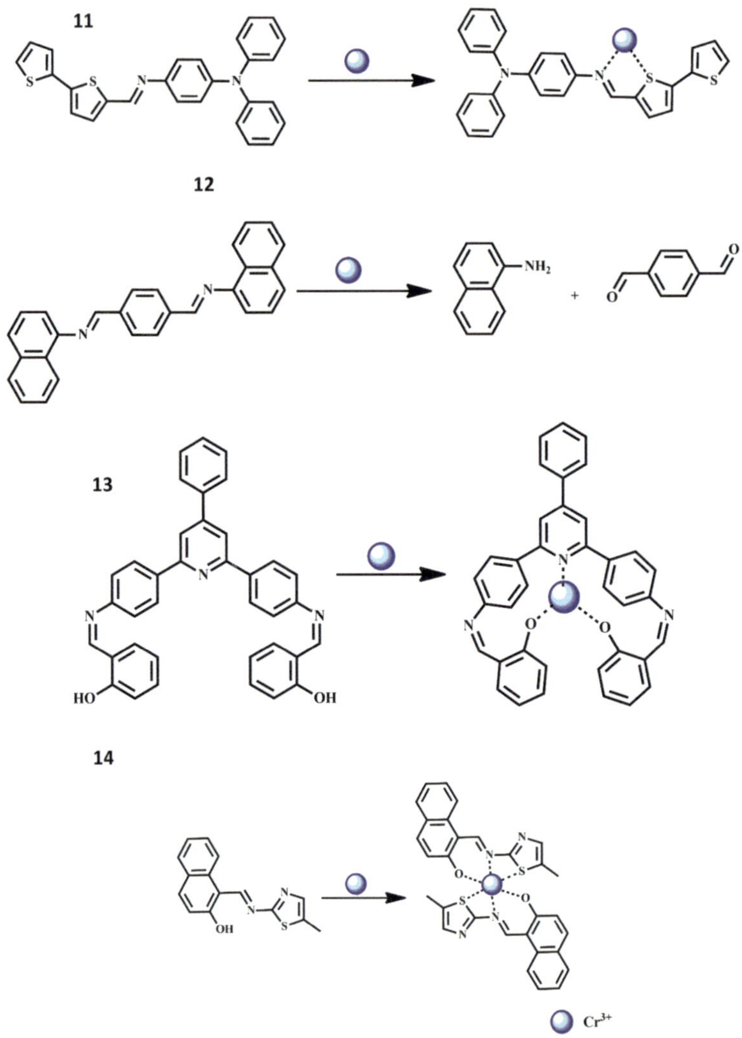

Figure 4.
Formation of Cr^{3+} complexes with Schiff base chemosensors 12–16.

Co^{2+}, Zn^{2+}, Pb^{2+}, Fe^{2+}, Fe^{3+}, Hg^{2+}, Ca^{2+}, Cd^{2+}, and Mn^{2+}), the chemosensor shows excellent selectivity toward As^{3+}. The development and evaluation of ArsenoFluor1, the first example of a chemosensor was done for the detection of As^{3+} in organic solvents at 298 K. At λem = 496 nm in THF, AF1 shows a 25-fold increase in fluorescence that is selective for As^{3+} over other physiologically relevant ions (such as Na^{+}, Mg^{2+}, Fe^{2+}, and Zn^{2+}) and has a sub-ppb detection limit. According to AF1's optical characteristics, a strong broad absorption band with a center wavelength of 385 nm in THF may be seen. This band is predominantly dominated by the coumarin chromophore. Due to effective quenching by the thiazoline N lone pair through a photoinduced electron transfer process, the resulting fluorescence emission maximum at 496 nm exhibits an incredibly low quantum yield (f) of 0.004. The nonconjugated AF1 is basically non-fluorescent. When As^{3+} is added (as AsI_3, but $AsCl_3$ also produces comparable effects), AF1's fluorescence intensity increases by around 25 times. This strong turn-ON response is accompanied by red shifts in the absorbance maxima from

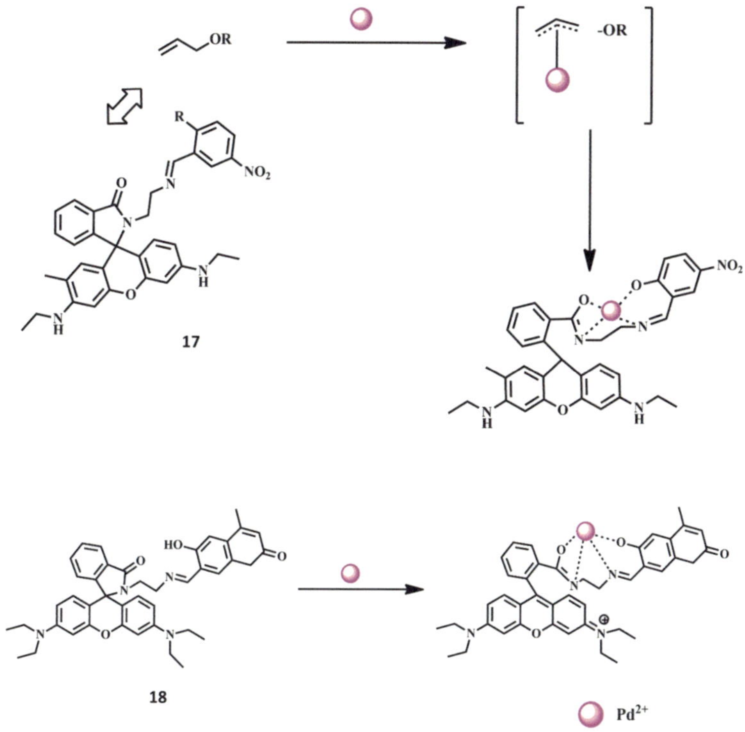

Figure 5.
Formation of Pd^{2+} complexes with Schiff base chemosensors 17–18.

385 to 464 nm, which are indicative of benzothiazole C6-CF3 production. The commercially available coumarin-6 (analogue of C6-CF3 with hydrogen replacing the CF3 group) and other coumarin-benzothiazole compounds typically exhibit fluorescence via an internal-charge-transfer (ICT) mechanism. As a result, at 298 K, AF1 functions as an efficient OFF–ON fluorescence sensor for As3+ in organic media. The sensing process probably entails bis-coordination of the Schiff-base thiolate form of AF1 to the thiolate anion, which then attacks the C-N carbon and loses a proton to produce the benzothiazole [58]. The structures of the chemosensors 19-21 and their interaction with Cr^{3+} ion are shown in **Figure 6**.

6. Detection of other cations

A fluorogenic and chromogenic chemosensor 22 with great sensitivity was created. The developed chemosensor was distinguished by contemporary analytical methods. The completely characterized chemosensor was utilized to detect the presence of Au^{3+} ions in acetonitrile-containing aqueous media in the presence of a variety of competing analytes, including cations (chlorides salts of Au^{3+}, Cr^{3+}, Fe^{3+}, Zn^{2+}, Cd^{2+}, Pb^{2+}, Hg^{2+}, Pt^{2+}, Pd^{2+}, perchlorates of Mn^{2+}, Co^{2+}, Ni^{2+}, Cu^{2+}, and nitrate salts of Ag^{+}, and Al^{3+}). The chemosensor 22 only demonstrated a selective response to Au^{3+} ions and revealed a 696-fold fluorescence increase toward Au^{3+}. The structure of the chemosensor 22 and its interaction with Au^{3+} ion are shown in **Figure 7**. The limit of detection was determined to be 1.51 M. The chemosensor 22 also had a colorimetric response that ranged from colorless to pink. These imine bond (C- -N)

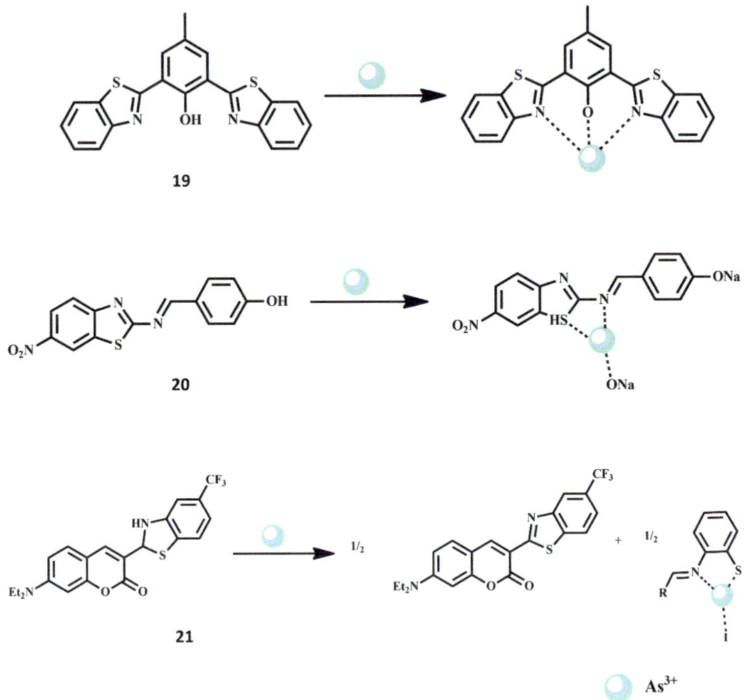

Figure 6.
Formation of As^{3+} complexes with Schiff base chemosensors 19–21.

hydrolysis-induced fluorometric and colorimetric responses were caused by Au^{3+} ions. Additionally, chemosensor 22 was successfully used to determine the presence of Au^{3+} ions in living MC3T3 cells [59]. For the purpose of detecting Y^{3+}, a new tetraphenylethylene decorated with m-aminobenzoic acid (chemosensor 23) was designed and produced with an 89% yield. Based on the AIE effect, it displayed the far-red fluorescence at 550–670 nm in aqueous environments. It demonstrated strong Y^{3+} selectivity across all types of metal ions and a clear "turn-on" fluorescence, which was seen for the first time for a Y^{3+} fluorescence sensor. The detection threshold for Y^{3+} was as low as 2.23×10^{-7} M. Fluorescence titration helped to clarify the sensor mechanism for a 1:1 stoichiometric ratio. The selective sensing of Y^{3+} was also effectively used to identify Y^{3+} for tested paper, a simulated water sample, and bio-imaging of live cells, indicating the strong practical application potential of this fluorescent sensor on detecting Y^{3+} in diverse water media and living body environments [60]. Similarly, a new fluorescent chemosensor 24 was architected for the recognition of Y^{3+} ions. The probe 2-hydroxy-1-naphtaldehyde salicyloylhydrazone was described in detail. For the determination of yttrium in THF, a very sensitive fluorescent approach and the fluorescence method were developed based on the chelation reaction; they can detect traces of Y^{3+} with a naked eye in sunlight [61]. The structures of the chemosensors 23–24 and their interactions with Y^{3+} ion are shown in **Figure** 7. For the detection of Ga^{3+}, a simple Schiff-base based on fluorene and salicylaldehyde was created. Using NMR titration, ESI-mass spectrometry analysis, photophysical experiments, and DFT simulations, the sensing behaviors of chemosensor 25 with Ga^{3+} were investigated. In the presence of a variety of cations, including K^{+}, Cu^{2+},

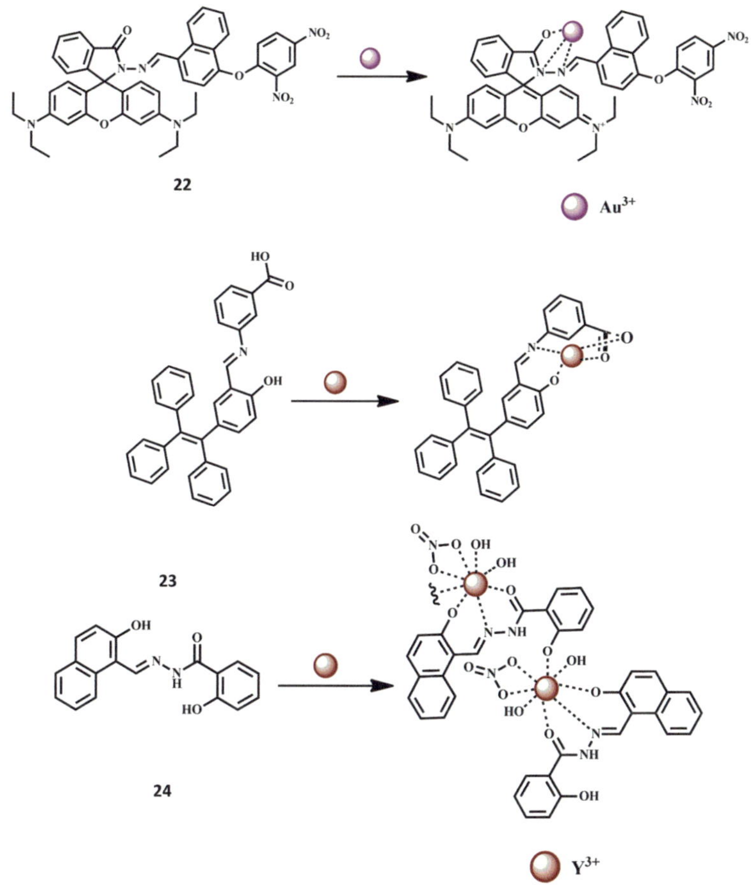

Figure 7.
Formation of Au^{3+} and Y^{3+} complexes with Schiff base chemosensors 22 and 23–24 respectively.

Co^{2+}, Ca^{2+}, Fe^{3+}, Cr^{3+}, Ag^{+}, Na^{+}, Ni^{2+}, In^{3+}, Fe^{2+}, Mg^{2+}, Cd^{2+}, Hg^{2+}, Mn^{2+}, Pb^{2+}, and Al^{3+}, the chemosensor 25 produced a highly stable complex and showed good selectivity for Ga^{3+}. In particular, the detection threshold for Ga^{3+} on chemosensor 25 was as low as 10 nM [62]. Chemosensor 26 and 27 were recently used for the effective detection of Ga^{3+} ions. The symmetrical Schiff base chemosensor 26 (N′2,N′5-bis((E)-2-hydroxybenzylidene)thiophene-2,5-dicarbohydrazide) and chemosensor 27 (N′5,N′7-bis((E)-2-hydroxybenzylidene)-2,3-dihydrothieno[3,4-b]) are both produced from thiophene. Fluorescence turn-on behavior was observed for Ga^{3+} detection. After chelation with Ga^{3+}, both of them demonstrated considerable fluorescence amplification in MeCN/H_2O buffer solution. The sensors, on the other hand, displayed ultralow detection limits of 3.90×10^{-9} M and 3.97×10^{-9} M, respectively. A 1:2 stoichiometry between the sensors and the Ga^{3+} was seen in the Job's plots, with association constants of 3.89×10^{9} M^2 and 8.59×10^{9} M^2, respectively. Theoretical investigations into molecule configuration, charge distribution, molecular orbitals, electronic transitions, and energy change results were produced in order to further explain the differences in sensing performance between the two sensors [63]. The structures of the chemosensors 25–27 and their interactions with Ga^{3+} ion are shown in **Figure 8**.

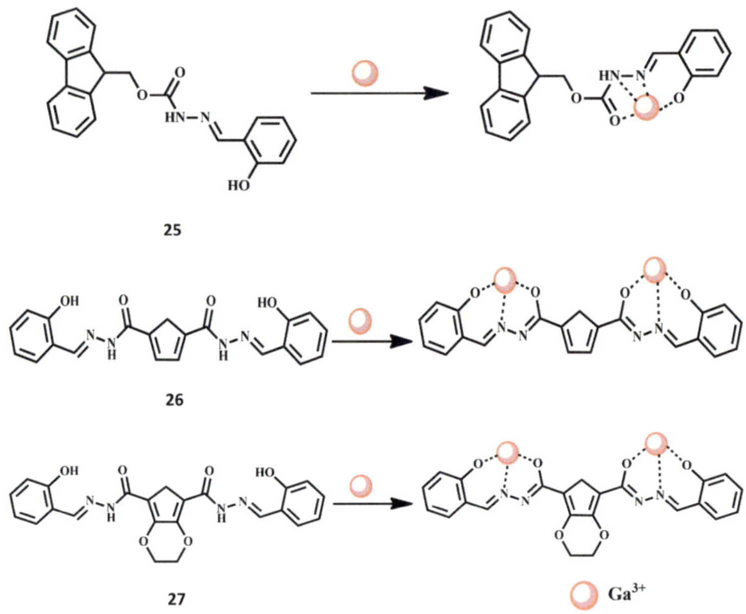

Figure 8.
Formation of Ga^{3+} complexes with Schiff base chemosensors 25–27.

7. Performance evaluation

We reviewed spectrofluorometric technique in this chapter for the identification and quantification of various cations in various environmental, biological, and agricultural materials. With a very low detection limit, many of these chemosensors have demonstrated remarkable sensitivity and selectivity for certain heavy metal analytes. The effectiveness of several Schiff base-based chemosensors for the detection of various cations is discussed. The best performing Schiff bases for sensing various cations were suggested after comparing the detection limit values. For the targeted detection of mercury ions, the chemosensors 1–6 were employed. With a detection limit of 8.3 nM, chemosensor 3 was discovered to be the most sensitive of these chemosensors. Cd^{2+} was detected using chemosensors 7–11. Fluorescence spectroscopy demonstrated the outstanding performance of chemosensor 7 for Cd^{2+}. The calculated detection threshold is 14.8 nM. Additionally, the use of chemosensor 7 for the detection of Cd^{2+} in real water samples was successful. Similar to this, when compared to chemosensors 12–16, chemosensor 13 had the best performance for Cr^{3+} detection. With a low detection limit of 0.94 nM, the chemosensor responded to Cr^{3+} by turning on its fluorescence. For the detection of Pd^{2+}, two different chemosensors using fluorescence (17–18) were employed. One of these chemosensors, 18 (with an outstanding detection limit of 0.0188 μM), was discovered to be very effective and sensitive for Pd^{2+}. Compared to chemosensors 19, 20, and 21 for the detection of As^{3+}, turn-on chemosensor 20 showed great sensitivity with a detection limit of 7.2 ppb in the aqueous medium. The Schiff base fluorescent chemosensor 22 displayed high sensitivity toward Au^{3+} with a detection limit of 1.51 μM, while the chemosensor 24 displayed good sensitivity toward Y^{3+} with a detection limit of 0.013 ppb. These findings come from the discussion on the sensing of Au^{3+}, Y^{3+}, and Ga^{3+}. A detection limit of 3.9 nm for the chemosensor 26 indicated that it had good sensitivity to Ga^{3+} [64].

The different chemosensors and its analytical parameters for the chemosensing analysis of heavy metals are listed in the **Table 1**.

8. Conclusion

Schiff base compounds are easily made in laboratories using various reaction procedures. With the help of various analytical techniques, Schiff bases have been used to detect a variety of metal ions, such as Hg^{2+}, Cd^{2+}, Cr^{3+}, Pd^{2+}, As^{3+}, Au^{3+}, Y^{3+}, and Ga^{3+}. This chapter has discussed how various Schiff base probes can be used as chemosensors to detect metal ions (cations). A stable complex can be formed when a metal center coordinates with the nitrogen atom of azomethine, which is a possible ligating site. In terms of metal cation detection, excellent results have been reported. However, the development of chemosensors with high selectivity and sensitivity for harmful metal cations that can operate in a variety of media is crucial. The availability, biological compatibility, and structural flexibility of Schiff base derivatives are further factors that support research.

Author details

Sabeel M. Basheer[1*], Puthiyavalappil Rasin[2], Vipin Manakkadan[2], Vishnunarayanan Namboothiri Vadakkedathu Palakkeezhillam[2] and Anandaram Sreekanth[2]

1 Department of Chemistry, VIT-AP University, Amaravati, Andhra Pradesh, India

2 Department of Chemistry, National Institute of Technology Tiruchirappalli, Tamil Nadu, India

*Address all correspondence to: sabeel.basheer@vitap.ac.in

References

[1] Fu F, Wang Q. Removal of heavy metal ions from wastewaters: A review. Journal of Environmental Management. 2011;**92**:407-418. DOI: 10.1016/j.jenvman.2010.11.011

[2] Li M, Gou H, Al-Ogaidi I, Wu N. Nanostructured sensors for detection of heavy metals: A review. ACS Sustainable Chemistry & Engineering. 2013;**1**:713-723. DOI: 10.1021/sc400019a

[3] Sharma N, Sodhi KK, Kumar M, Singh DK. Heavy metal pollution: Insights into chromium eco-toxicity and recent advancement in its remediation. Environmental Nanotechnology, Monitoring & Management. 2021;**15**:100388. DOI: 10.1016/j.enmm.2020.100388

[4] Briffa J, Sinagra E, Blundell R. Heavy metal pollution in the environment and their toxicological effects on humans. Heliyon. 2020;**6**:04691. DOI: 10.1016/j.heliyon.2020.e04691

[5] Lentini P, Zanoli L, Granata A, Signorelli SS, Castellino P, Dell'Aquila R. Kidney and heavy metals - The role of environmental exposure (review). Molecular Medicine Reports. 2017;**15**:3413-3419. DOI: 10.3892/mmr.2017.6389

[6] Fu Z, Xi S. The effects of heavy metals on human metabolism. Toxicology Mechanisms and Methods. 2020;**30**:167-176. DOI: 10.1080/15376516.2019.1701594

[7] Muro-Gonzalez DA, Mussali-Galante P, Valencia-Cuevas L, Flores-Trujillo K, Tovar-Sanchez E. Morphological, physiological, and genotoxic effects of heavy metal bioaccumulation in Prosopis laevigata reveal its potential for phytoremediation. Environmental Science and Pollution Research. 2020;**27**:40187-40204. DOI: 10.1007/s11356-020-10026-5

[8] Anyanwu BO, Orisakwe OE. Current mechanistic perspectives on male reproductive toxicity induced by heavy metals. Journal of Environmental Science and Health, Part C: Environmental Carcinogenesis and Ecotoxicology Reviews. 2020;**38**:204-244. DOI: 10.1080/26896583.2020.1782116

[9] Vardhan KH, Kumar PS, Panda RC. A review on heavy metal pollution, toxicity and remedial measures: current trends and future perspectives. Journal of Molecular Liquids. 2019;**290**:111197. DOI: 10.1016/j.molliq.2019.111197

[10] Igbokwe IO, Igwenagu E, Igbokwe NA. Aluminium toxicosis: A review of toxic actions and effects. Interdisciplinary Toxicology. 2020;**12**:45-70. DOI: 10.2478/intox-2019-0007

[11] Wu X, Cobbina SJ, Mao G, Xu H, Zhang Z, Yang L. A review of toxicity and mechanisms of individual and mixtures of heavy metals in the environment. Environmental Science and Pollution Research. 2016;**23**:8244-8259. DOI: 10.1007/s11356-016-6333-x

[12] Zhou Q, Lei M, Liu Y, Wu Y, Yuan Y. Simultaneous determination of cadmium, lead and mercury ions at trace level by magnetic solid phase extraction with Fe@ Ag@Dimercaptobenzene coupled to high performance liquid chromatography. Talanta. 2017;**175**:194-199. DOI: 10.1016/j.talanta.2017.07.043

[13] Bansod BK, Kumar T, Thakur R, Rana S, Singh I. A review on various electrochemical techniques for heavy metal ions detection with different

sensing platforms. Biosensors & Bioelectronics. 2017;**94**:443-455. DOI: 10.1016/j.bios.2017.03.031

[14] Zinoubi K, Majdoub H, Barhoumi H, Boufi S, Jaffrezic-Renault N. Determination of trace heavy metal ions by anodic stripping voltammetry using nanofibrillated cellulose modified electrode. Journal of Electroanalytical Chemistry. 2017;**799**:70-77. DOI: 10.1016/j.jelechem.2017.05.039

[15] Colim AN, do Nascimento PC, Wiethan BA, Adolfo FR, Dresch LC, de Carvalho LM, et al. Reversed-phase High-performance Liquid Chromatography for the Determination of 15 Rare Earth Elements in Surface Water Sample Collected in a Mining Area from Lavras do Sul/RS. Vol. 82. Brazil: Chromatographia; 2019. pp. 843-856. DOI: 10.1007/s10337-019-03709-w

[16] Shih TT, Hsu IH, Chen SN, Chen PH, Deng MJ, Chen Y, et al. A dipole-assisted solid-phase extraction microchip combined with inductively coupled plasma-mass spectrometry for online determination of trace heavy metals in natural water. Analyst. 2015;**140**:600-608. DOI: 10.1039/c4an01421a

[17] Comby S, Tuck SA, Truman LK, Kotova O, Gunnlaugsson T. New trick for an old ligand! The sensing of Zn(II) using a lanthanide based ternary Yb(III)-cyclen-8-hydroxyquinoline system as a dual emissive probe for displacement assay. Inorganic Chemistry. 2012;**51**:10158-10168. DOI: 10.1021/IC300697W

[18] Aletti AB, Gillen DM, Gunnlaugsson T. Luminescent/colorimetric probes and (chemo-) sensors for detecting anions based on transition and lanthanide ion receptor/binding complexes. Coordination Chemistry Reviews. 2018;**354**:98-120. DOI: 10.1016/J.CCR.2017.06.020

[19] Gunnlaugsson T, Bichell B, Nolan C. A novel fluorescent photoinduced electron transfer (PET) sensor for lithium. Tetrahedron Letters. 2002;**43**:4989-4992. DOI: 10.1016/S0040-4039(02)00895-X

[20] Di W, Sedgwick AC, Thorfinnur Gunnlaugsson EU, Akkaya JY, James TD. Fluorescent chemosensors: The past, present and future. Chemical Society Reviews. 2017;**46**:7105-7123. DOI: 10.1039/C7CS00240H

[21] Tidwell TT. Hugo (Ugo) Schiff, Schiff bases, and a century of β-lactam synthesis. Angewandte Chemie, International Edition. 2008;**47**:1016-1020. DOI: 10.1002/anie.200702965

[22] De La Hoz A, Díaz-Ortiz A, Prieto P. R.S.C. Green Chem. In: Microwave-assisted Green Organic Synthesis. UK: Royal Society of Chemistry; 2016. pp. 1-33. DOI: 10.1039/9781782623632-00001

[23] Liu YT, Sheng J, Yin DW, Xin H, Yang XM, Qiao QY, et al. Ferrocenyl chalcone-based Schiff bases and their metal complexes: Highly efficient, solvent-free synthesis, characterization, biological research. Journal of Organometallic Chemistry. 2018;**856**:27-33. DOI: 10.1016/j.jorganchem.2017.12.022

[24] Naeimi H, Salimi F, Rabiei K. Mild and convenient one pot synthesis of Schiff bases in the presence of P2O5/Al2O3 as new catalyst under solvent-free conditions. Journal of Molecular Catalysis A: Chemical. 2006;**260**:100-104. DOI: 10.1016/j.molcata.2006.06.055

[25] Kwon MS, Kim S, Park S, Bosco W, Chidrala RK, Park J. One-pot synthesis of imines and secondary amines by

Pd-catalyzed coupling of benzyl alcohols and primary amines. The Journal of Organic Chemistry. 2009;**74**:2877-2879. DOI: 10.1021/jo8026609

[26] Khan E, Khan S, Gul Z, Muhammad M. Medicinal importance, coordination hemistry with selected metals (Cu, Ag, Au) and chemosensing of thiourea derivatives. A review. Critical Reviews in Analytical Chemistry. 2020;**50**:1-23. DOI: 10.1080/10408347.2020.1777523

[27] Kaur B, Kaur N, Kumar S. Colorimetric metal ion sensors – A comprehensive review of the years 2011-2016. Coordination Chemistry Reviews. 2018;**358**:13-69. DOI: 10.1016/j.ccr.2017.12.00

[28] Gunnlaugsson T, Ali HDP, Glynn M, Kruger PE, Hussey GM, Pfeffer FM, et al. Journal of Fluorescence. 2005;**15**:287. DOI: 10.1007/s10895-005-2627-y

[29] de Silva AP, Gunaratne HQN, Gunnlaugsson T, Huxley AJM, McCoy CP, Rademacher JT, et al. Chemical Reviews. 1997;**97**:1515-1566. DOI: 10.1021/cr960386p

[30] Rotkiewicz K, Grellmann KH, Grabowski ZR. Reinterpretation of the anomalous fluorescense of p-n,n-dimethylamino-benzonitrile. Chemical Physics Letters. 1973;**19**:315-318. DOI: 10.1016/0009-2614(73)80367-7

[31] Rettig W. Photophysical and photochemical switches based on twisted intramolecular charge transfer (TICT) states. Applied Physics B Photophysics and Laser Chemistry. 1988;**45**:145-149. DOI: 10.1007/BF00695283

[32] Ito A, Ishizaka S, Kitamura N. A ratiometric TICT-type dual fluorescent sensor for an amino acid. Physical Chemistry Chemical Physics. 2010;**12**:6641-6649. DOI: 10.1039/B924176K

[33] Rettig W, Lapouyade R. Fluorescence probes based on twisted intramolecular charge transfer (TICT) states and other adiabatic photoreactions. In: Lakowicz JR, editor. Topics in Fluorescence Spectroscopy, Probe Design and Chemical Sensing. Vol. 4. New York: Plenum Press; 1994. p. 109. DOI: 10.1007/0-306-47060-8_5

[34] Lodeiro C, Pina F. Luminescent and chromogenic molecular probes based on polyamines and related compounds. Coordination Chemistry Reviews. 2009;**253**:1353. DOI: 10.1016/j.ccr.2008.09.008

[35] Wu JS, Zhou JH, Wang PF, Zhang XH, Wu SK. New fluorescent chemosensor based on exciplex signaling mechanism. Organic Letters. 2005;7:2133. DOI: 10.1021/ol0504648

[36] Chen WB, Elfeky SA, Nonne Y, Male L, Ahmed K, Amiable C, et al. A pyridinium cation–π interaction sensor for the fluorescent detection of alkyl halides. Chemical Communications. 2011;**47**:253. DOI: 10.1039/C0CC01420F

[37] Shellaiah M, Rajan YC, Balu P, Murugan A. A Pyrene based schiff base probe for selective fluorescent turn-on detection of Hg^{2+} ions with live cell application. New Journal of Chemistry. 2015;**39**:2523-2531. DOI: 10.1039/c4nj02367f

[38] Wei T, Gao G, Qu W, Shi B, Lin Q, Yao H, et al. Selective fluorescent sensor for mercury(II) ion based on an easy to prepare double naphthalene Schiff base. Sensors and Actuators B: Chemical. 2014;**199**:142-147. DOI: 10.1016/j.snb.2014.03.084

[39] Muthusamy S, Rajalakshmi K, Zhu D, Zhu W, Wang S, Lee K-B, et al. Dual detection of mercury (II) and lead (II) ions using a facile coumarin-based

fluorescent probe via excited state intramolecular proton transfer and photo-induced electron transfer processes. Sensors and Actuators B: Chemical. 2021;**346**:130534. DOI: 10.1016/j.snb.2021.130534

[40] İnal EK. A Fluorescent chemosensor based on schiff base for the determination of Zn^{2+}, Cd^{2+} and Hg^{2+}. Journal of Fluorescence. 2020;**30**:891-900. DOI: 10.1007/s10895-020-02563-6

[41] Su Q, Niu Q, Sun T, Li T. A simple fluorescence turn-on chemosensor based on Schiff-base for Hg2+−selective detection. Tetrahedron Letters. 2016;**57**(38):4297-4301. DOI: 10.1016/j.tetlet.2016.08.031

[42] Wang XM, Yan H, Chen Y, Bao HB. A new Schiff's base fluorescent sensor for the selective detection of mercury ion. Advanced Materials Research. 2011;**239-242**:1105-1108. DOI: 10.4028/www.scientific.net/AMR.239-242.1105

[43] Mohanasundaram D, Bhaskar R, Gangatharan Vinoth Kumar G, Rajesh J, Rajagopal G. A quinoline based Schiff base as a turn-on fluorescence chemosensor for selective and robust detection of Cd^{2+} ion in semi-aqueous medium. Microchemical Journal. 2021;**164**:106030. DOI: 10.1016/j.microc.2021.106030

[44] Ma J, Dong Y, Zhou Y. Yan Wub and Zhen Zhao A pyridine based Schiff base as a selective and sensitive fluorescent probe for cadmium ions with "turn-on" fluorescence responses. New Journal of Chemistry. 2022;**46**:3348. DOI: 10.1039/D1NJ05919J

[45] Maniyazagan M, Mariadasse R, Jeyakanthan J, Lokanath NK, Naveen S, Premkumar K, et al. Rhodamine based "turn–on" molecular switch FRET–sensor for cadmium and sulfide ions and live cell imaging study. Sensors and Actuators B: Chemical. 2017;**238**:565-577. DOI: 10.1016/j.snb.2016.07.102

[46] Aydin Z. A turn-on fluorescent sensor for cadmium ion detection in aqueous solutions. JOTCSA. 2020;**7**(1):277-286. DOI: 10.18596/jotcsa.638912

[47] Kolcu F, Erdener D, Kaya İ. A Schiff base based on triphenylamine and thiophene moieties as a fluorescent sensor for Cr (III) ions: Synthesis, characterization and fluorescent applications. Inorganica Chimica Acta. 2020:119676. DOI: 10.1016/j.ica.2020.119676

[48] Ding Y, Zhao C. A highly selective fluorescent sensor for detection of trivalent metal ions based on a simple schiff-base. Química Nova. 2018. DOI: 10.21577/0100-4042.20170229

[49] Chalmardi GB, Tajbakhsh M, Hasani N, Bekhradnia A. A new Schiff-base as fluorescent chemosensor for selective detection of Cr 3+: An experimental and theoretical study. Tetrahedron. 2018;**74**(18):2251-2260. DOI: 10.1016/j.tet.2018.03.046

[50] Khan S, Muhammad M, Algethami JS, Al-Saidi HM, Almahri A, Hassanian AA. Synthesis, characterization and applications of schiff base chemosensor for determination of Cr(III) Ions. Journal of Fluorescence. 2022;**32**(5):1889-1898. DOI: 10.1007/s10895-022-02990-7

[51] Abbasi A, Shakir M. An inner filter effect based Schiff base chemosensor for recognition of Cr(vi) and ascorbic acid in water matrices. New Journal of Chemistry. 2018;**42**(1):293-300. DOI: 10.1039/c7nj03677a

[52] Bhanja A, Mishra S, Saha K, Sinha C. A fluorescence 'turn-on'chemodosimeter

for the specific detection of Pd $^{2+}$ by a rhodamine appended Schiff base and its application in live cell imaging. Dalton Transactions. 2017;**46**:9245-9252. DOI: 10.1039/C7DT01288H

[53] Adak AK, Purkait R, Manna SK, Ghosh BC, Pathak S, Sinha C. Fluorescence sensing and intracellular imaging of Pd2+ ions by a novel coumarinyl-rhodamine Schiff base. New Journal of Chemistry. 2019;**43**:3899-3906. DOI: 10.1039/c8nj06511j

[54] Martinez VD, Vucic EA, Becker-Santos DD, Gil L, Lam WL. Arsenic exposure and the induction of human cancers. Journal of Toxicology. 2011;**2011**:1-13. DOI: 10.1155/2011/431287

[55] Yu H-S, Liao W-T, Chai C-Y. Arsenic carcinogenesis in the skin. Journal of Biomedical Science. 2006;**13**(5):657-666. DOI: 10.1007/s11373-006-9092-8

[56] Hong Y-S, Song K-H, Chung J-Y. Health effects of chronic arsenic exposure. Journal of Preventive Medicine and Public Health. 2014;**47**(5):245-252. DOI: 10.3961/jpmph.14.035

[57] Hu S, Yan G, Wu C, He S. An ethanol vapor sensor based on a microfiber with a quantum-dot gel coating. Sensors. 2019;**19**(2):300. DOI: 10.3390/s19020300

[58] Shah AQ, Kazi TG, Baig JA, Arain MB, Afridi HI, Kandhro GA, et al. Determination of inorganic arsenic species (As^{3+} and As^{5+}) in muscle tissues of fish species by electrothermal atomic absorption spectrometry (ETAAS). Food Chemistry. 2010;**119**(2):840-844. DOI: 10.1016/j.foodchem.2009.08.0

[59] Paul S, Bhuyan S, Mukhopadhyay SK, Murmu NC, Banerjee P. Sensitive and selective in vitro recognition of biologically toxic As(III) by rhodamine based chemoreceptor. ACS Sustainable Chemistry & Engineering. 2019. DOI: 10.1021/acssuschemeng.9b00935

[60] Memon S, Bhatti AA, Bhatti AA, Ocak Ü, Ocak M. A new calix[4]arene Schiff base sensor for Hg^{2+} and Au^{3+}. Journal of the Iranian Chemical Society. 2016;**13**(12):2275-2282. DOI: 10.1007/s13738-016-0946-3

[61] Jiang S, Chen S, Guo H, Yang F. "Turn-on" far-red fluorescence sensor for Y3+ based on Schiff-based tetraphenylethylene. Dyes and Pigments. 2020;**183**:108717. DOI: 10.1016/j.dyepig.2020.108717

[62] Zhang D, Zang Z, Zhou X, Zhou Y, Tang X, Wei R, et al. A selective fluorescence probe for yttrium(III) based on acylhydrazone Schiff base. Inorganic Chemistry Communications. 2009;**12**(11):1154-1156. DOI: 10.1016/j.inoche.2009.08.007

[63] Seong YL, Kwon HB, Tae GJ, So YK, Cheal K. A simple Schiff-base fluorescence probe for the simultaneous detection of Ga^{3+} and Zn^{2+}. Inorganica Chimica Acta. 2017;**461**:127-135. DOI: 10.1016/j.ica.2017.02.019

[64] Xing Y, Liu Z, Li B, Li L, Yang X, Zhang G. The contrastive study of two thiophene-derived symmetrical Schiff bases as fluorescence sensors for Ga^{3+} detection. Sensors and Actuators B: Chemical. 2021;**347**:130497. DOI: 10.1016/j.snb.2021.130497

Chapter 9

Schiff Base Complexes for Catalytic Application

Shakeel Nawaz Shariff, Supriya Saravu and Dileep Ramakrishna

Abstract

Primary amines are combined with an aldehyde group to generate Schiff base compounds, which are called condensation imine products. This class of compounds has a general structure, R-C=NR', where R and R' represent alkyl/aryl/cyclohexyl/heterocyclic group. These compounds contain an azomethine group that is basic in nature due to, (i) the presence of lone pair of electrons on the nitrogen and (ii) electron-donating nature of the double bond. Hence, these compounds, as ligands, participate in the formation of metal complexes. The presence of lone pair of electrons on the nitrogen atom and the hybridization involved explains the physical, chemical, and spectral properties of nitrogen-containing moieties. In the case of (sp^2) hybridization (trigonal structure), the lone pair of electrons occupies either a symmetrical unhybridized 2p orbital that is perpendicular to the plane of trigonal hybrids or a symmetrical hybrid orbital, whose axis is in the plane, leaving behind only the π-electrons in the unhybridized 2p orbital. A very similar type of hybridization is experienced by the nitrogen atom in the azomethine group. Traditional phosphine complexes of nickel, palladium, and platinum, particularly those of palladium, have played an extremely important role in the development of homogeneous catalysis. Schiff base complexes as catalysts have been studied for various organic transformations such as oxidation, epoxidation, reduction, coupling reactions, polymerization reactions, hydroformylations, and many more.

Keywords: Schiff base metal-complexes, homogenous catalysis, oxidation, C-C coupling, epoxidation

1. Introduction

Schiff base ligands (SBLs) are known as "privileged ligands" [1] due to their facile synthesis by the condensation of aldehydes with imines. In 1864, Hugo Schiff first introduced the condensation of an aldehyde and an amine, which leads to the formation of a Schiff base (SB) [2]. SBLs are capable to form coordination bonds with metals *via* imine nitrogen and group, which is generally connected to the aldehyde. In recent days, Schiff bases (SBs) are prepared where efficient and well-synthesized SBLs are labeled as "privileged ligands." SBs can stabilize various metals with different oxidation states and control the behavior of metals in diverse catalytic transformations, which are beneficial to most industries [3]. SBLs are easily synthesized and can combine with practically any metal ion to form complexes [4]. At high temperatures (>100°C) and with moisture, several Schiff base complexes (SBCs) exhibit strong catalytic activity in various organic reactions. There have been numerous studies in recent years explaining application in catalysis; therefore, this chapter highlights the catalytic activity of SBCs. Almost all SBCs exhibit strong catalytic activity [4]. SBs are often bi- or tridentate ligands that can bind to transition metals to produce incredibly stable complexes. In some cases, liquid crystals are also employed for this purpose as most liquid crystals have an azomethine group. The formation of carbon-nitrogen bonds can be accomplished through Schiff base reactions in organic synthesis [5]. Coordination compounds were once thought to be an infrequent and unique group, and later they were acknowledged as the most important compounds, which are capable to transform straightforward inorganic molecules to form organic matter. To give a good example of a coordination component, chlorophyll (the magnesium porphyrin complex found in plants) converts CO_2 and H_2O into carbohydrates.

The basic structural formula for SBs is RRC=N-R, where R, R', here R is labeled as hydroxyalkyl, alkyl, cyclohexyl, hydroxyaryl, etc. SBs consist of azomethine moiety (C=N) an azomethine group. In some cases, -SH or –OH functional groups also participate in complexation, when they are enough close to azomethine moiety. This is due to the lone pair of electron's being donated to the electron deficient metal ion. Overall, this facilitates the development of a metal complex with good stability. With salicylideneimine, Ettling created the copper (II) complex in 1840. Since then, there has been a significant quantity of material published on SBs and related metal complexes (MCs) [6].

2. Synthetic routes to Schiff base complexes

From the past two decades, there is significant progress in the formation of SB transition MCs, with an increased focus on synthesis and structure. The characteristic of the SB and metal ion determines the synthesis, characteristics, and structure of the SBCs. Mostly, the following techniques have been employed to create MCs of SBs [7].

Aldehydes and amines can condense with one another under various reaction circumstances in the presence of various solvents. Methanol and ethanol are the typical solvents used to prepare the SBs, either at room temperature or under refluxing conditions. SBs are typically more likely to form when dehydrating chemicals such as magnesium sulfate are present. If the syntheses are conducted in organic solvents, the water generated during the reaction can be extracted easily. During the purification process, the SBs may deteriorate. It is preferable in these circumstances to crystallize the SBs and purify them. Here, in a suitable solvent, SB is permitted to interact with metal ions. The use of a binary azeotropic mixture of water and an organic solvent has been documented in some situations, and using an organic solvent is recommended to prevent the hydrolysis of the azomethine group.

2.1 Direct synthesis

Here, a suitable solvent is used to facilitate the reaction between SB and metal ions. Despite the fact that there have been instances of mixing organic solvent and water, the use of organic solvent is recommended to prevent the hydrolysis of the azomethine group [8].

2.2 *In situ* method

This procedure offers the addition of the metal ion during or immediately after the aldehyde and amine have reacted with each other. Metal ions have been found to typically pair with just one element. Before other components are added, the reaction speeds up toward the formation of the complex [9].

2.3 Coordinated secondary amine oxidation

For this kind, the secondary amine's oxidative dehydrogenation is facilitated by the metal ion followed by the direct formation of Schiff base complexes. In **Figure 1**, one such response is provided.

2.4 Amine exchange reaction

This process involves the reaction of an amine and a transition metal complex of SBs, which results in the exchange of aryl and alkyl groups as given below [10].

$$Ni(o\text{-}OC_6H_4CH{=}NR_2) + 2R'NH_2 \rightarrow Ni(o\text{-}OC_6H_4CH{=}NR_2) + 2RNH_2.$$

Where: R = n-C_4H_9, H, o-HSC_6H_4, C_4H_{10}, C_6H_5,

2.5 Metal exchange reaction

This is an interesting route for the preparation of simple metal to carbon bond. The sample below shows a common metal interchange reaction [10].

$$CH_3(CH_2)_3\text{-}Li + KOC(CH_3)_3 \rightarrow CH_3(CH_2)_3\text{-}K + LiOC(CH_3)_3$$

2.6 Ligand exchange reaction

It is very challenging to synthesize molecules such as dioxomolybdenum(VI) complexes, for which there are different techniques. The most preferred method is the ligand exchange method. Below is an example of a common response [11].

$$MoO_2(acac)_2 + SalenH_2 \rightarrow MoO_2(Salen) + 2acacH$$

Where: acac = acetylacetone.

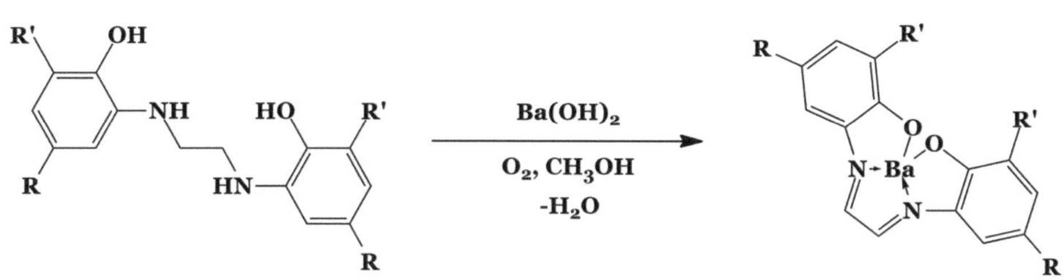

Where R=H, n-C4H9, C6H5, R'=C4H9, C6H5, o-HSC6H4

Figure 1.
Synthesis of SB from a secondary amine.

2.7 Template synthesis

A template reaction in coordination chemistry is any of a group of ligand-based reactions that take place between two or more nearby coordination sites on a metal center. When there is no metal ion present, the same organic reactants result in different end products. Although the coordination modifies the electronic properties of ligands (acidity, electrophilicity, etc.), the template effects highlight the pre-organization provided by the coordination sphere. In a general sense, transition metal-based catalysis can be viewed as template reactions: Reactants coordinate to adjacent sites on the metal ion and, owing to their adjacency, the two reactants interconnect (insert or couple) either directly or *via* the action of another reagent. In the area of homogeneous catalysis, the cyclo-oligomerization of acetylene to cyclooctatetraene at a nickel(II) center reflects the templating effect of the nickel, where it is supposed that four acetylene molecules occupy four sites around the metal and react simultaneously to give the product. The development of these catalytic reactions was influenced by these crude mechanistic hypotheses. For instance, only three acetylene molecules could bind if a competing ligand, such as triphenylphosphine, was added to occupy one coordination site. These three molecules then combine to form benzene.

This process involves the interaction of an aldehyde, an amine, and a metal compound in a single step to create complexes without the need to isolate Schiff bases [12, 13]. Metal ions are serving as a catalyst for these reactions. Rotaxanes, helicates, macro cycles, and catenanes are just a few examples of assemblies with unique topologies that have been produced using template synthesis [14]. The information needed to arrange a group of building blocks so that they can be connected in a certain way is therefore said to be contained in a templating agent. Thermodynamic and kinetic processes are the two different categories of these templates. The template binds to one of the reactants in the first case, shifting the equilibrium in favor of the product formation. With regard to kinetic processes, the templates work under conditions that are irreversible, stabilizing each transition state and causing the formation of the intended product. The template is tightly coupled to the end species in many kinetically regulated processes. In these situations, it serves as both a thermodynamic template and a kinetic template. Gimeno et al. defined a template can be class that arranges a grouping of molecular construction pieces through noncovalent interactions that favor the creation of a particular complex in their review (**Figure 2**) [15].

2.8 Rearrangement of heterocycles (thiazole and oxazoles)

When an *o*-mercapto amine or o-amino, and o-hydroxy are directly combined with a carbonyl molecule to generate a SB, ring closure and the production of a heterocyclic compound are frequently the unintended side reactions [16]. Hugo Schiff first converted SB into its metal chelate in 1869 [17]. According to reports, the benzothiazoline and benzoxazole rings can open in particular situations when metal ions are present, and this ring rearrangement creates the necessary metal chelates for the SBs. This type of ring opening complexation process is shown in **Figure 3**.

3. Chemistry of Schiff bases

Hugo Schiff, a Nobel Prize laureate, first identified SBs as a condensation outcome of carbonyl compounds and primary amines [18]. SB is a structural counterpart of an –

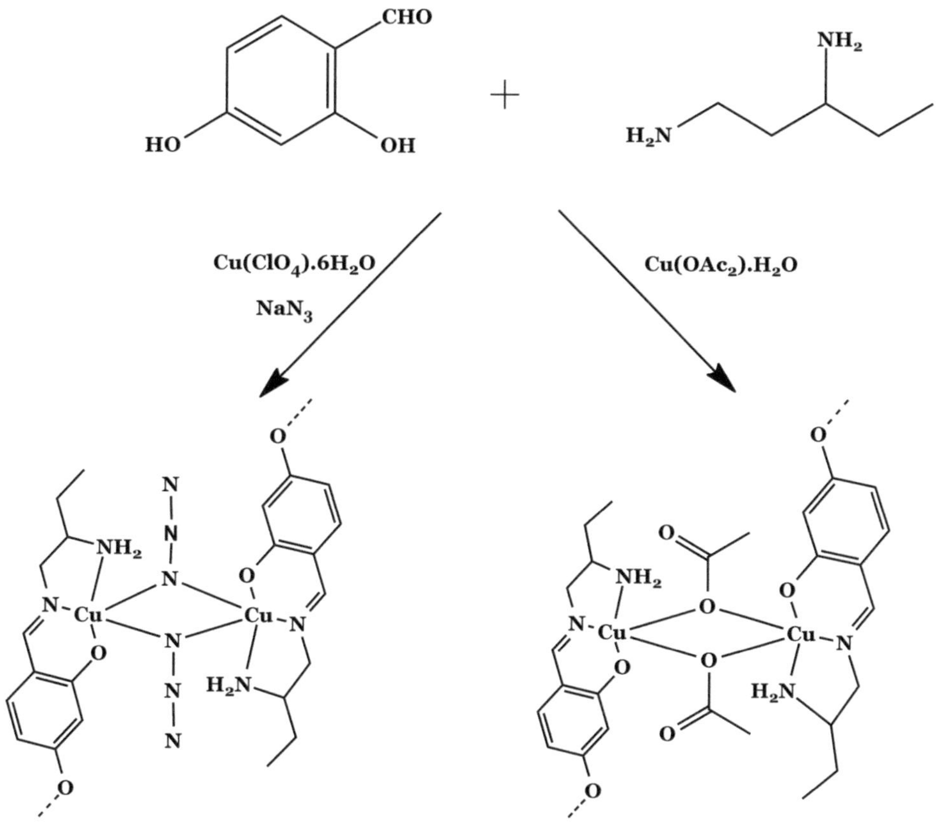

Figure 2.
Bridged dinuclear copper (II) SB complex using template method.

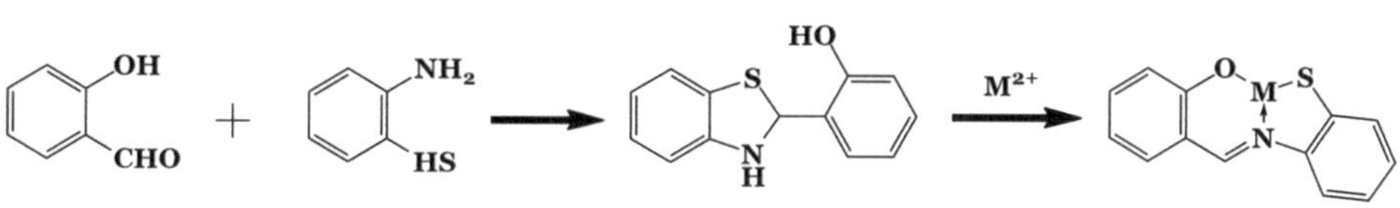

Figure 3.
2-(2-hydroxyphenyl) benzothiazoline.

CHO or –C=O where C=O has been exchanged out for an imine or an azomethine group (**Figure 4**) [3, 19].

A chemical molecule known as a Schiff Base, or SB, contains a C-N double bond, where the N is linked to an aryl group or an alkyl (R) but, not to an H-atom. A molecule containing azomethine linkage is the same as the SB. Hugo Schiff was honored by having his name attached to several compounds, which generally have the following structure:

The SB is an imine since R stands for an alkyl or phenyl group, making it a stable compound. Even the imine N and other group, typically connected with aldehyde, this type of ligands, can coordinate metal ions. SBs are still created by chemists, and today's "privileged ligands" are active, well-designed Schiff base ligands [20]. The bridging SBs possess below given structure, where it consists of several functional groups that can be altered depending on the situation (**Figure 5**).

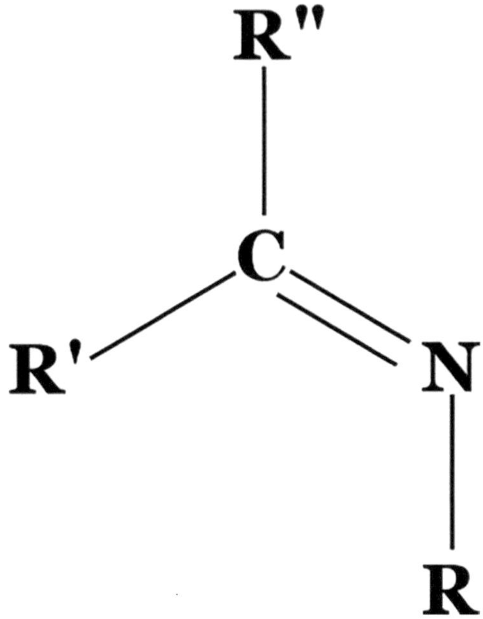

Figure 4.
Schiff base.

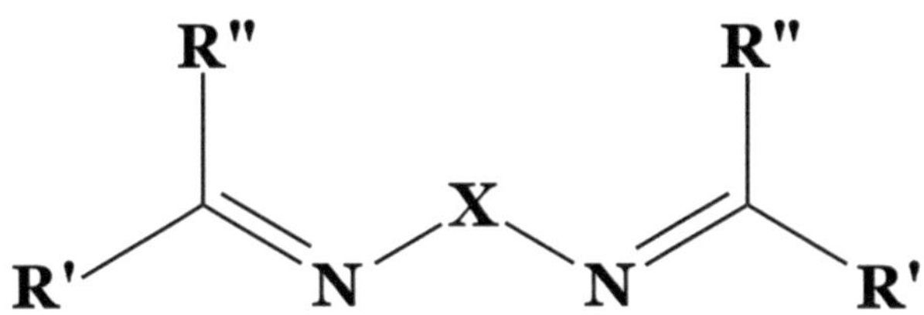

Figure 5.
Bridged SB.

Where R' here might be a H or R group, R" is a phenyl or substituted phenyl, and X is alkyl or aryl group. In fact, SBs can influence the activity of metals in a broad range of beneficial reactions in catalysis by stabilizing a wide variety of metals in a variety of oxidation states [21]. The oxygen atoms in SBs are often donated by NO or N_2O_2, although they can be substituted by S, N, or Se atoms. **Figure 6** shows the synthesis of SB by condensation reaction.

R, here, might be either aryl or an alkyl group. Aryl SBs are significantly more stable and easier to synthesize than their alkyl counterparts, which are more prone to instability. Aliphatic aldehyde SBs are comparatively not stable and easily undergo polymerization [22]. Aldehydes or ketones can be converted into SBs in a reversible process that often occurs when the substance is heated or when an acid or a base catalyzes the reaction (**Figure 7**).

In most cases, the product is separated from the reaction, or when the water formed is removed, or sometimes in both cases. Aqueous acid or base has the ability to hydrolyze several Schiff bases back to their respective aldehydes, ketones, and amines. Nucleophilic addition to the carbonyl group is a common theme in the synthetic route of Schiff base formation. The amine serves as a nucleophile during this

Figure 6.
Synthesis of SB by condensation reaction.

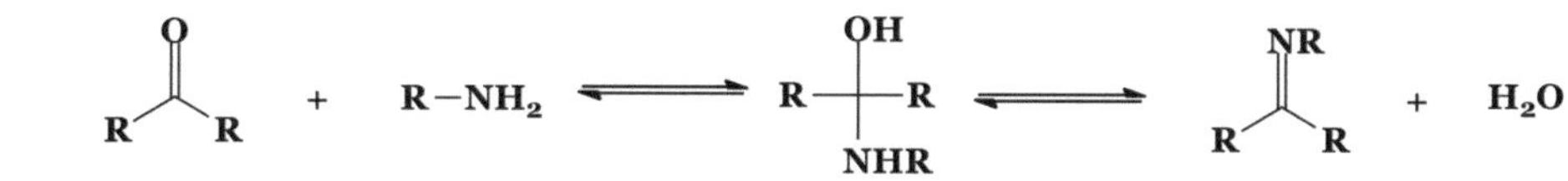

Figure 7.
Reversible reaction of SB formed by the interaction of aldehydes or ketones.

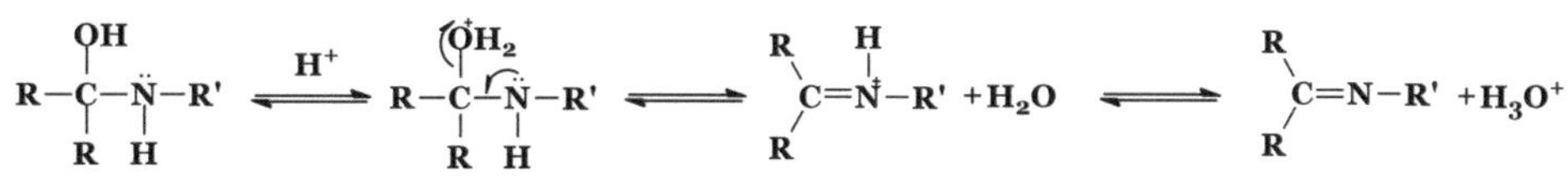

Figure 8.
Mechanism of formation SB.

instance. In the first step of the procedure, the amine reacts with the aldehyde or ketone to produce an unstable addition substance known as carbinolamine. Either an acid- or a base-catalyzed pathway results in the carbinolamine losing water. This carbinolamine undergoes acid-catalyzed dehydration since it is an alcohol (**Figure 8**).

Usually, the rate-determining step in the creation of a SB is dehydration of the carbinolamine, which is why acids are used to catalyze the reaction. Yet because amines are basic chemicals, the acid concentration should not be exceeded. Equilibrium shifts to the left, and carbinolamine production is prevented if the amine is protonated and turns non-neutral. As a result, it is ideal to synthesize numerous Schiff bases at a pH that is a little acidic. The base also acts as a catalyst for the dehydration of carbinolamines. It goes through an anionic intermediate in two phases. In reality, the SB creation is a series of two different kinds of reactions, namely addition and elimination [23].

4. SB complexes in catalysis

Excellent catalytic activity is displayed by several SBCs in diverse reactions and in the presence of moisture. Numerous studies on their use in homogeneous and heterogeneous catalysis have been published during the past few years. The high heat and moisture stabilities of many Schiff base complexes proved helpful for their application as catalysts in high-temperature processes. Due to the fact that complexation often causes an increase in activity, knowledge of the characteristics of both ligands and metals can help in the synthesis of substances that have high activity [14].

SBs were treated as crucial ligands in coordination chemistry for a long time [24, 25]. SBs are an essential type of ligands in coordination chemistry [26–28]. Owing to their relatively straightforward synthesis and diverse structural makeup, SBs and their metal complexes were extensively studied because of their extraordinary properties and uses in variety of fields. These compounds are revealed to be auspicious for the synthetic and structural research field [1, 29, 30].

From the finding of chiral Mg (III) salen SB catalysts in the asymmetric epoxidation of unfunctionalized olefins in the 1990s, the tetradentate SB, salen-type ligand, and their complexes have attracted significant attention [31]. Due to their potential application as catalysts in a variety of C-C and C-N bond formation events, such as cyclopropanations, aziridination, asymmetric cycloaddition reactions, and A3-coupling, among others, Salen-Schiff base transition MCs have been well studied (**Figure 9** [32, 33]. For medicinal chemists, the design of the C-N bonds of aryl compounds is extremely crucial. Tetrazoles are a significant group of heterocycles among the many heterocycles that have been reported. In coordination chemistry, tetrazoles have been utilized as ligands, [34] stabilizers in the photographic industry, [35] linkers to covalently bind synthetic groups to biopolymers in a specific manner, etc. [36]. The A^3-coupling reaction is an appealing illustration of a multi-component reaction [37]. (**Figure 10**) [38]. Agrahari and coworkers carried out an A^3-coupling reaction with less use of catalyst, and the catalyst was reclaimed for four successive cycles in the reactions [39].

The primary method for producing biaryls is the Suzuki-Miyaura cross-coupling reaction. This reaction is an effective technique for the production of medicines due to the properties and convenient accessibility of organoboron reagents [40]. Palladium is one of the transition metals that are most frequently employed in both industry and academia to produce C-N, C-S, and C-C bonds because it is a

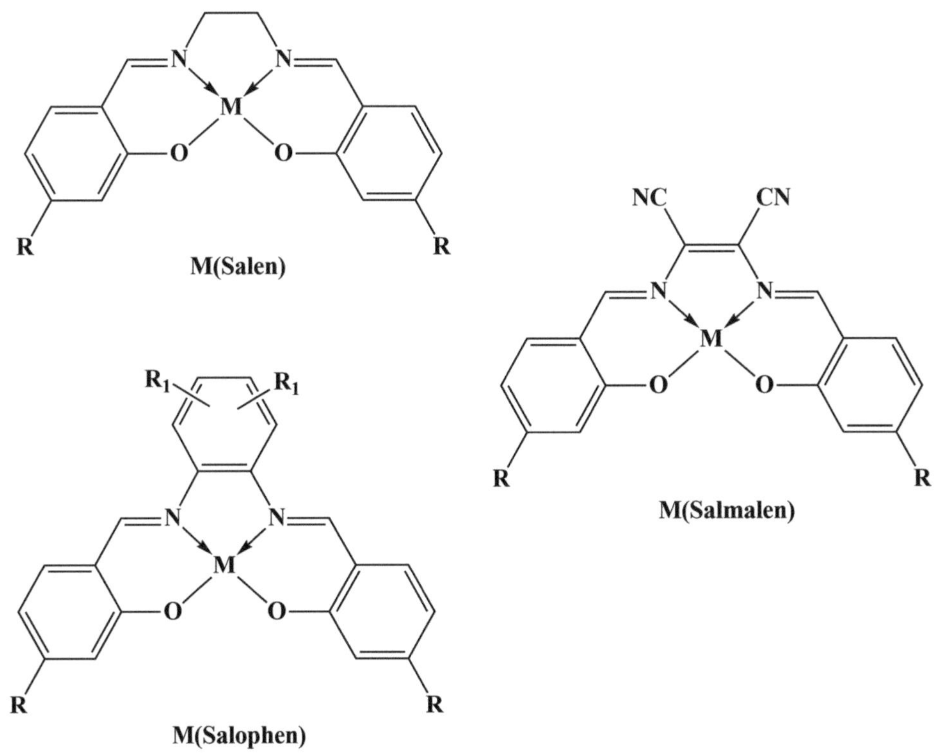

Figure 9.
Metal salen-type complexes.

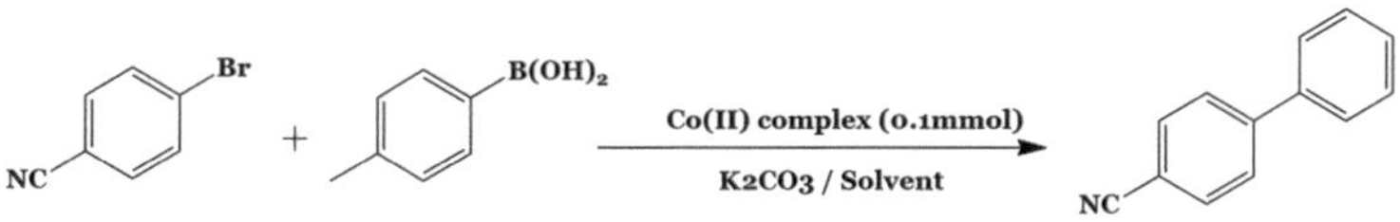

Figure 10.
4-bromobenzonitrile with phenylboranic acid.

catalytically active metal [41]. In this context, during the past few decades, significant attempts have been made to synthesize potential substitute catalysts for cross-coupling reactions [42]. Transition metals in the first row are a particularly promising class of molecules that have the potential to be more useful in the field of catalysis than other transition MCs [43]. This work deals with the less harmful alternative to the commonly utilized palladium-phosphine ligand complexes for initiating Suzuki reactions: SB-Cobalt complexes. This work describes the synthesis of OON- and ONN-SBLs and their Co complexes. The created compounds were evaluated as Suzuki coupling reaction catalysts. This work utilized only 0.1 mM catalyst for the reaction [44].

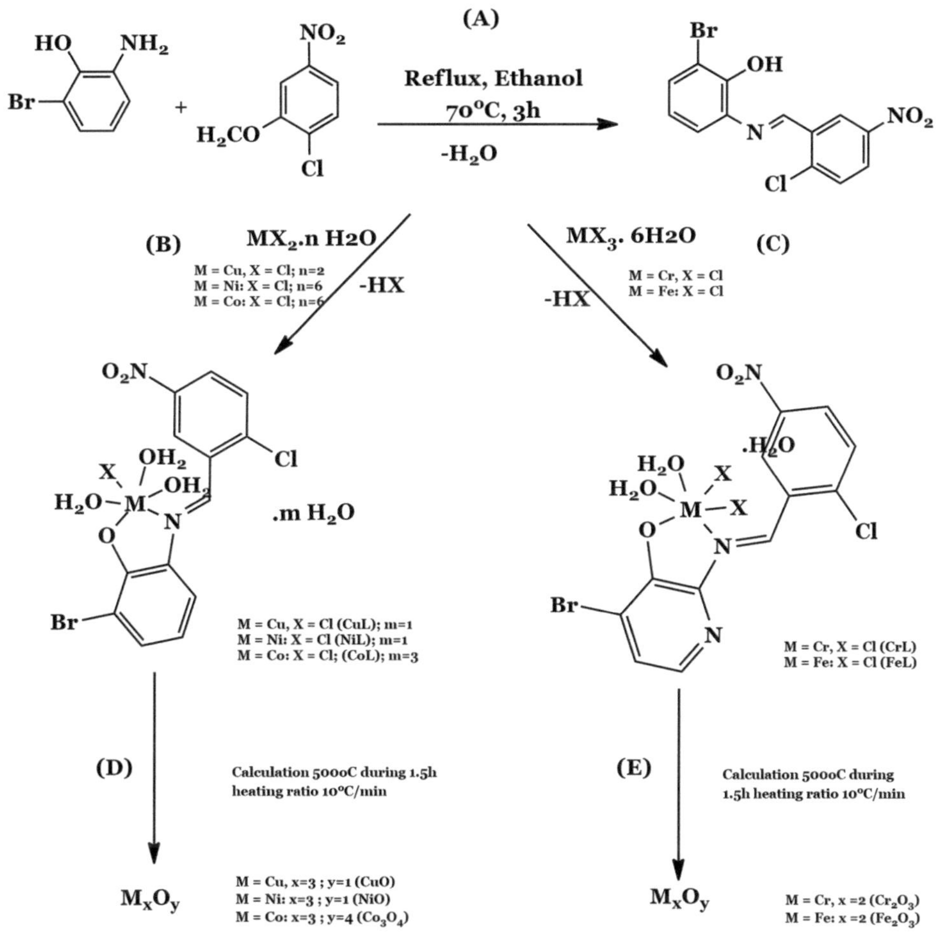

Figure 11.
Scheme 1 (a–c) synthetic method for the preparation of Schiff base complex of BCNPOH and metal oxides.

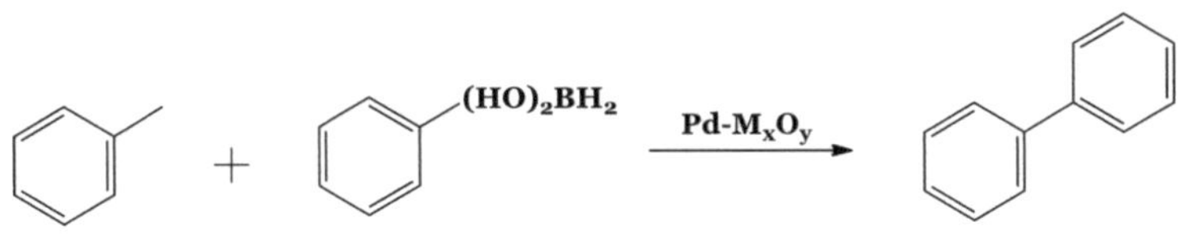

Figure 12.
Scheme3: C-C coupling reaction of phenylboronic acid and iodobenzene.

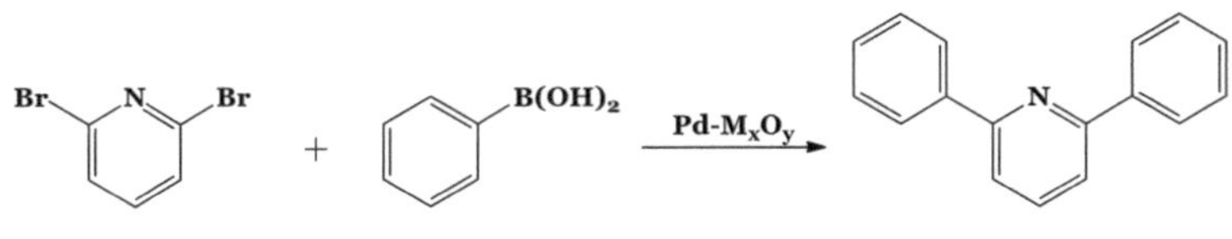

Figure 13.
C-C coupling reaction of phenylboronic acid and 2, 6-dibromopyridine.

SBL, BCNPOH, and metal salts (Cu, Co, Ni, Fe, and Cr) were individually dissolved in ethanol and followed by mixing both the solutions and reflux for 5 h at 70°C. The obtained precipitate was filtered and dried. CuL, CoL, NiL (**Figure 10**, Scheme 1b), FeL, and CrL (**Figure 11**, Scheme 1c) produced SB complexes that were recrystallized in ethanol [45].

The synthesized complexes were implemented for the C–C coupling reaction of Phenylboronic Acid and Iodobenzene and Phenylboronic Acid and 2,6-Dibromopyridine.

The coupling reaction between iodobenzene and phenylboronic acid was conducted (**Figure 12**). The reaction was completed in 95:5 DMF:water mixture under an inert nitrogen environment.

The coupling reaction between 2,6-dibromopyridine and phenylboronic acid was assessed at 110°C for 24 hours. (**Figure 13**). The reaction was taken to react in 5 mL of 95:5 DMF:water combination under an inert nitrogen environment with the limiting reagent 2,6-dibromopyridine and a molar ratio of the halide to the boronic acid of 1:2.4. The amount of catalyst was always 25 mg, and the molar ratio of the base (K_2CO_3) to the halide was 1:2 [45].

5. Synthesis for the transfer hydrogenation of ketones

Through the use of pre-catalyst Ru(II) (**Figure 14**), the catalytic transfer hydrogenations of ketones were investigated (0.001 mmol). In *i*-propanol, the base KOH (4.0 mmol) was dissolved (5 mL). This combination was then combined with 2.0 mmol of the substrate ketone, and it was refluxed in an oil bath at 80°C for 8 hours. After the reaction was finished, the liquid was cooled and filtered using alumina or silica gel, and then eluted using an hexane /ethyl acetate (4,1) mixture. Compound purity was evaluated using GCMS. All of the complexes (1a–d) were discovered to be efficient catalysts, offering conversion rates between 70 and 100%. The transfer hydrogenations of acetophenone derivatives were successfully accelerated by a variety of unique Ru(II) complexes that were successfully synthesized in the current work and contained pyridine group-based Schiff bases. With benzophenone as a model substrate, the efficiency of the catalysts in the transfer hydrogenations was evaluated (**Figure 15**) [46].

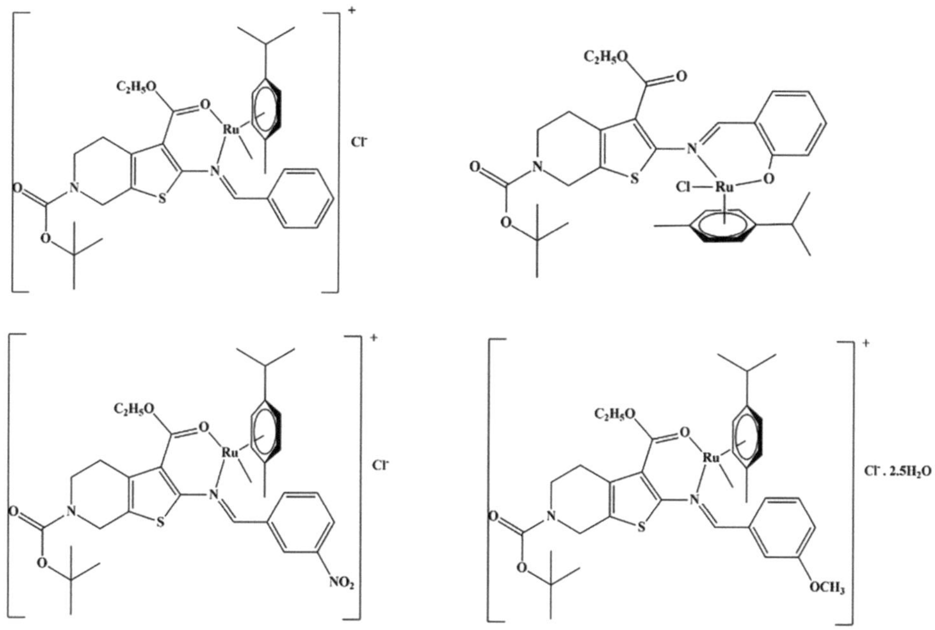

Figure 14.
Structures of Ru(II) pre-catalyst.

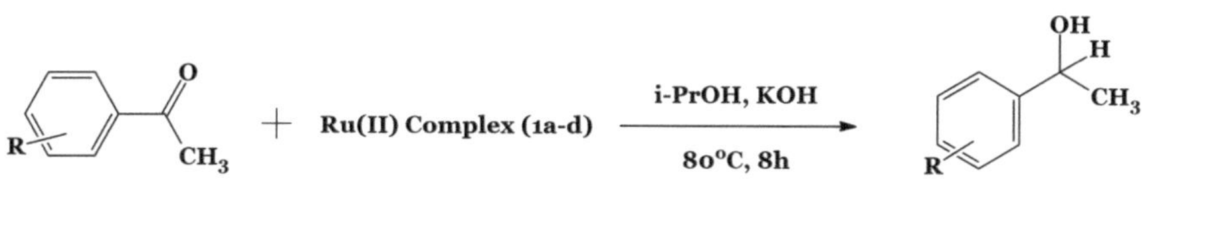

R= Br, H, OCH_3, Ph

Reaction condition; Substract (2.0 mmol), Ru(II) complex (0.001 mmol), KOH (4.0 mmol), i-PrOH (5ml), 8h, 80°C

Figure 15.
Hydrogenation of ketone using Ru(II).

Reliable zinc complexes for bulk polymerization of lactide have been studied at 150°C under applicable commercial circumstances. The utilized anionic Schiff base ligands (**Figure 14**) may be produced aerobically and are extremely stable against air, moisture, and other lactide contaminants. As a result, these compounds incorporate both a strong anionic ligand and a nontoxic, active zinc core (**Figure 16**). These compounds are ideally suited for lactide polymerization under industrially relevant circumstances due to their simple aerobic synthesis, high thermal stability under nitrogen, and air [47].

Cu-SBCs based on particularly ordered mesoporous silica MCM-41 were synthesized by Niakan et al., to create a heterogeneous catalyst (**Figure 17**). The synthesized catalyst was implemented for coupling reactions of Ullmann type. The synthesized catalyst showed excellent activity for the N-arylation of amines with aryl iodides and bromides. Here, 0.8 mol% of catalyst was used [48].

Bunno and group synthesized metal-containing SB/ Sulfoxides that served as chiral ligands for asymmetric intramolecular allylic C-H amination processes that are

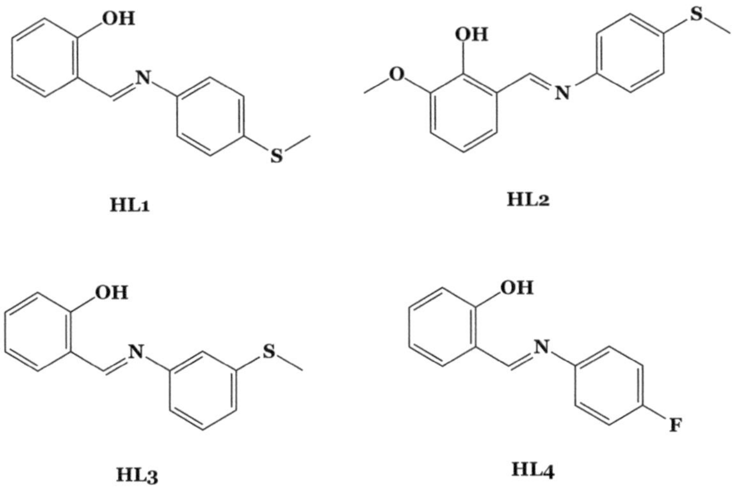

Figure 16.
SBLs HL1–HL4, used for the synthesis of four new zinc Schiff base complexes.

catalyzed by Pd(II).They claimed that the use of metal-containing Schiff base ligands helped to tune the reaction conditions. In the reaction, internal and terminal alkenes were both utilized [49]. A novel Cu(II) SBC on graphene oxide, GO-SB-Cu was successfully synthesized through surface functionalization of the GO (**Figure 18**). Potential of the catalyst was tested as heterogeneous catalyst in one-pot three component click reactions for synthesis of 2H-indazoles and 1,4-disubstituted 1,2,3-triazoles (**Figure 19**).

A reaction was carried out with aniline and 2-bromobenzaldehyde in the presence of sodium azide. Different reactions were carried out to set the optimized conditions using catalyst loading, solvent selectivity, temperature, and time for catalytic study of GO-SB [50].

One more important catalytic conversion reaction is in oxidation of alcohols into carbonyl compounds. A cobalt (II)-SBC with triphenylphosphine was synthesized (**Figure 20**). The catalyst was used for oxidation reactions using different substrates and obtained good % of conversion of carbonyl compound [51]. Dileep et al. have done experiment on the oxidation of primary and secondary alcohols to respective carbonyl compounds *via* green catalytic approach. Here, $[NiCl_2(PPh_3)_2]$ was synthesized by the reaction between $NiCl_2.6H_2O$ and triphenylphosphine in the presence of glacial acetic acid. Condensation reactions between 2-hydrazinopyridine and substituted salicylaldehyde in methanol yielded Schiff bases with 70–80%. This study examines the relationship between catalyst concentration and substrate. According to this study, 0.02 mmol of catalyst was enough to effectively convert benzyl alcohol to benzaldehyde [52]. Novel palladium complexes were synthesized and characterized by different analytical techniques (**Figure 21**). The complexes exhibit strong catalytic activity when N-methyl-morpholine-N-oxide is present as an oxidant for the oxidation of primary and secondary alcohols. A mechanism of catalytic conversion reaction is shown below (**Figure 22**) [53].

A novel Ru-SBC composed of Ph_3P was synthesized by and characterized by various techniques. The synthesized complexes were implemented for the oxidation of alcohols in [EMIM]Cl ionic liquid-NaOCl system. Recycling of catalyst is an important factor in catalysis. Herein, recyclability of catalyst for the conversion of

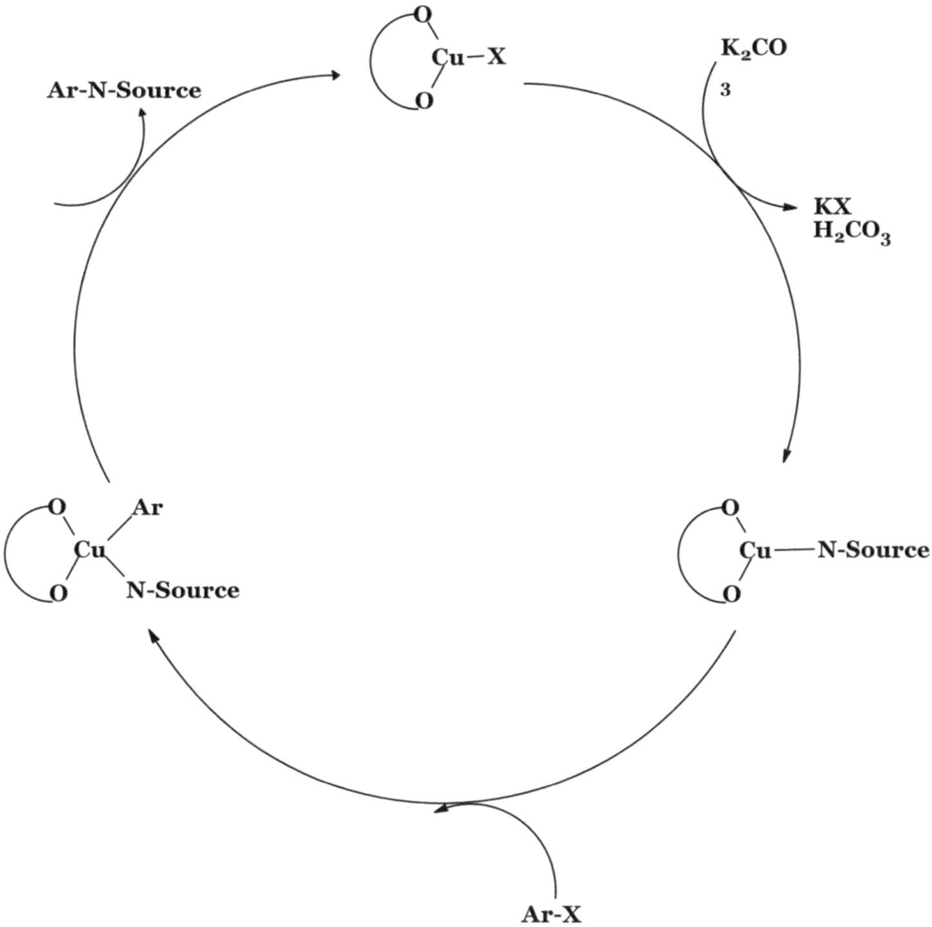

Figure 17.
A plausible mechanism for the Cu-SB-MCM-41-catalyzed C-N bond formation reaction.

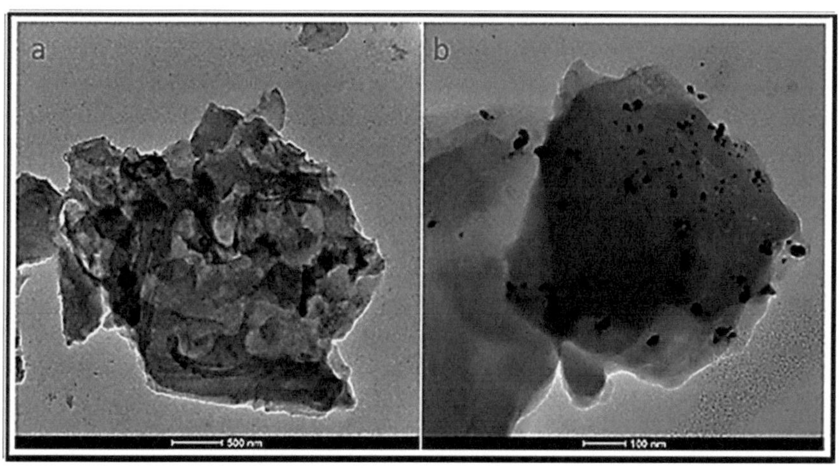

Figure 18.
HRTEM images of GO-SB-Cu.

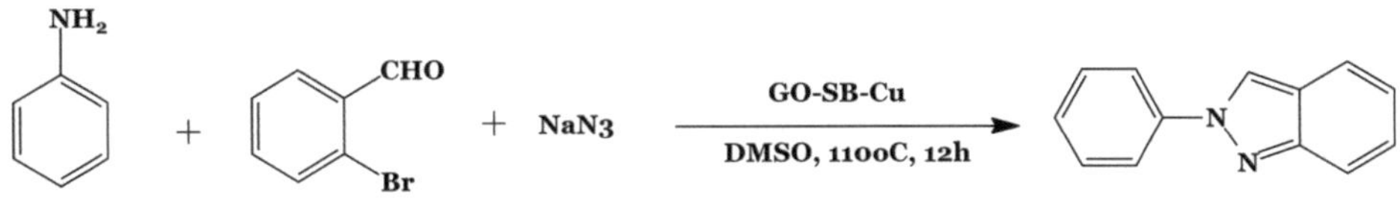

Figure 19.
Synthesis of 2H-indazole.

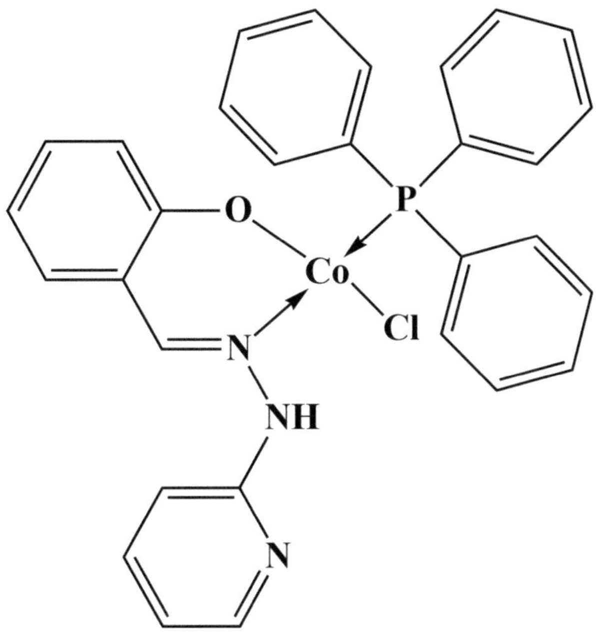

Figure 20.
Cobalt (II)-SBC with triphenylphosphine.

benzylalcohol is carried till seven consecutive cycles, shown in **Figure 23**. Recyclability graph indicates that recycling of ionic liquid has small significant effect on the conversion of the product [54].

For the catalytic oxidation of primary and secondary alcohols with periodic acid (H_5IO_6) as the oxidant, novel copper complexes were created and put into use [55]. Palladium was used as a catalyst in the carbonylation of alcohols in ionic liquid (IL) media (1-ethyl-3-methylimidazolium hexafluorophosphate). The usage of Pd-based catalyst, where NaOCl is utilized as an oxidant, considerably improved the carbonylation of primary/secondary alcohols to aldehydes/ketones [56]. *In situ* strategy of immobilization of a Cu complex onto an ionic liquid support was focused in this work. Olefin and terpene epoxidation was carried out by the immobilization of a copper complex and 1-ethyl-3-methylimidazoliumhexafluorophosphate, along with H_2O_2 as the terminal oxidant is developed. This method brings a clean environment nature for catalytic epoxidations (**Figure 24**) [57].

A SB Pd complex was studied for the catalytic activity in ethyl-methyl imidazolium hexafluorophosphate [EMIM] PF_6 ionic liquid. This work shows air stable system for Heck and Suzuki reactions. In the [EMIM] PF_6-Water system, the palladium-catalyzed coupling process demonstrated remarkable fidelity [58]. Copper complex

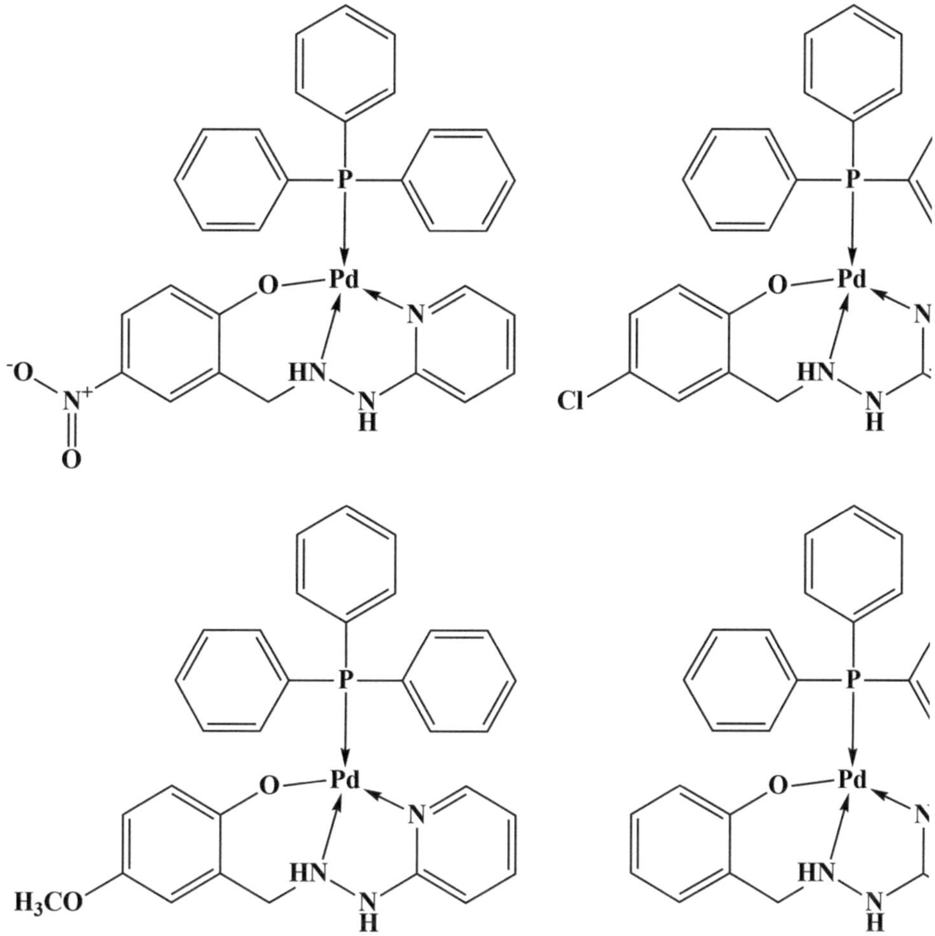

Figure 21.
Palladium SBC with triphenylphosphine.

was studied for the catalytic application in Heck and Suzuki reactions (**Figure 25**). The reactions were conducted in [EMIM]PF_6 ionic liquid-water mixture with 1:1 ratio, which was observed to be helpful during catalyst recycling and reusing [59]. These compounds demonstrated improved catalysis for the N-alkylation step required to produce benzazoles (**Figure 26**). The reaction progressed *via* a hydrogen transfer mechanism (**Figure 27**). It was observed that the reaction in EMIM [PF_6]-Water [1] mixture as solvent bought a considerably better yield with in a less reaction time (**Figure 28**) [60].

6. Conclusions

Schiff's base properties have been demonstrated to play an important role in many responses as a key to increasing yield and product selection. Important structural kinds and Schiff base ligand types were discovered by the catalytic activity of Schiff base complexes, such as the early oligomerization of isoprene and butadienes. Schiff base and OH group have demonstrated greater olefin epoxidation effectiveness than

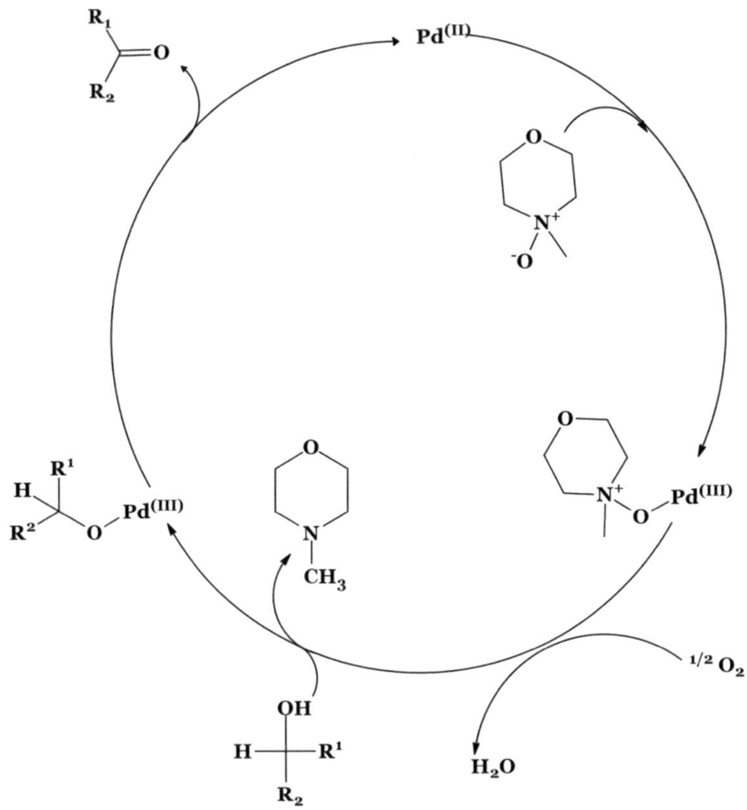

Figure 22.
Proposed mechanism for palladium complex catalyzed oxidation of alcohols by NMO.

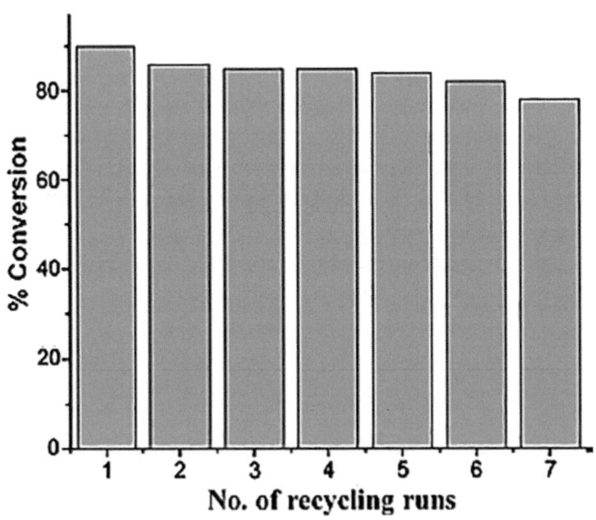

Figure 23.
Results of recycling on conversion of benzylalcohol.

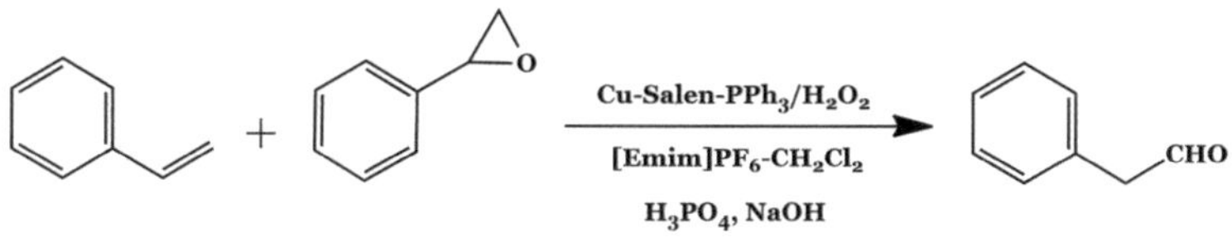

Figure 24.
Scheme for epoxidation of alkenes.

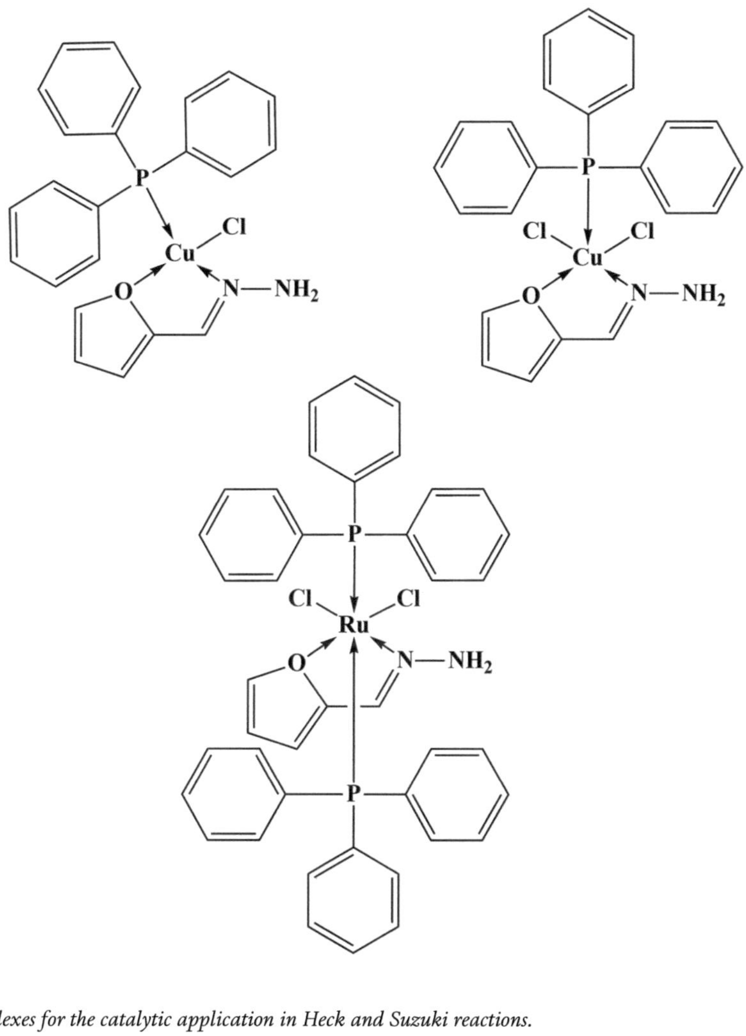

Figure 25.
Metal complexes for the catalytic application in Heck and Suzuki reactions.

untreated ligands or aryl. As detailed in this chapter, the flexible metals are combined with Schiff base ligands to assure a good function in changing the reaction to industrial value. Over the years, the structure of the metal structures containing the Schiff base has piqued the public's interest. Because of their low cost and ease of synthesis, as well as their chemical and thermal stability, Schiff base transition metal complexes are a popular oxidation catalyst for a wide range of organic substrates. Important reactions catalyzed by Schiff base metal complexes include oxidation reactions,

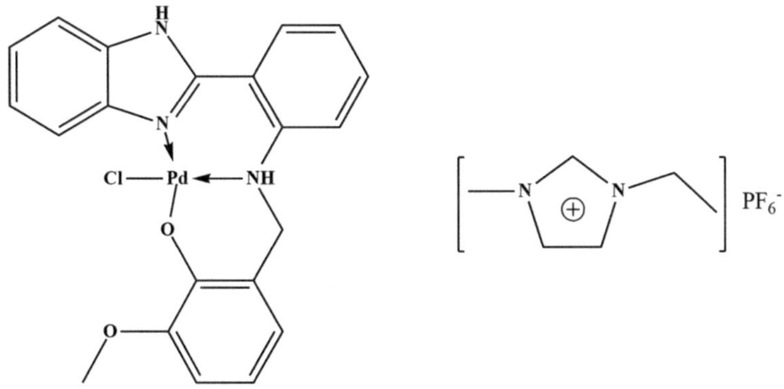

1 2

Figure 26.
Structures of benzimidazole-based Schiff base palladium complex (1) and [(EMIM) PF_6 (2).

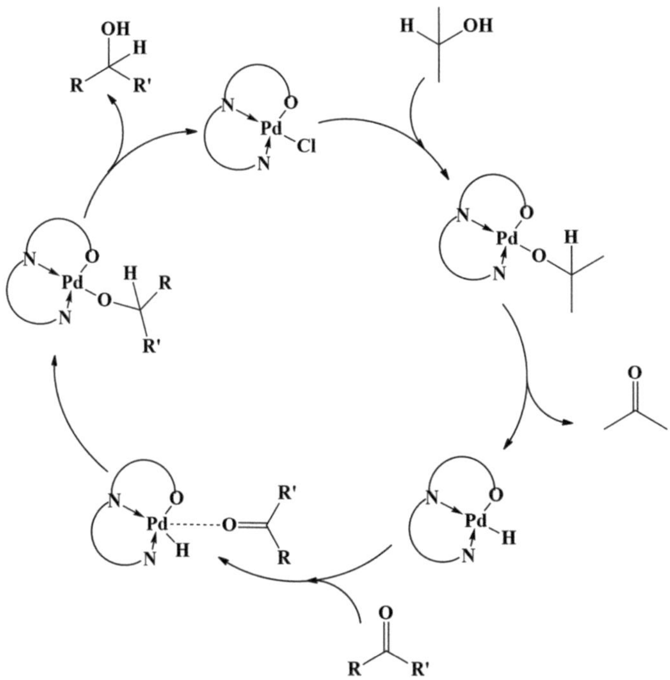

Figure 27.
Mechanism for the palladium complex catalyzed transfer hydrogenation.

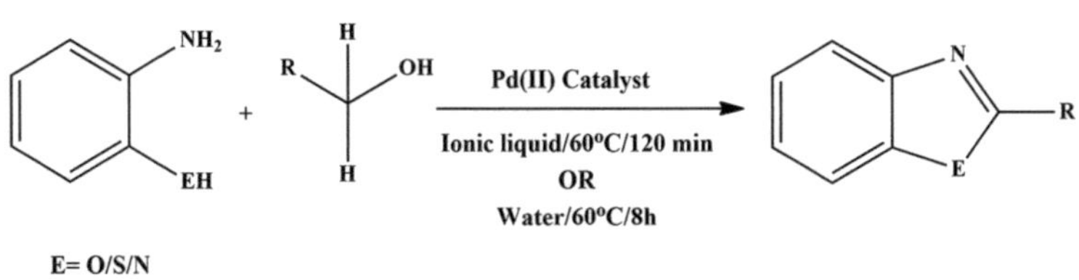

Figure 28.
Synthesis of benzazoles.

reductions, polymerizations, hydroformylation, coupling reactions, the oxidation of sulfides to sulfoxides, alkenes to epoxides and diols, and the activation of hydrocarbons. Yet there is still a huge need for new Schiff bases and their metal complexes as catalysts for organic transformations.

Author details

Shakeel Nawaz Shariff[1], Supriya Saravu[2] and Dileep Ramakrishna[1*]

1 Department of Chemistry, Presidency University, Bengaluru, India

2 Department of Chemistry, BMS College of Engineering, Bengaluru, India

*Address all correspondence to: dileep.618@gmail.com;
dileep.r@presidencyuniversity.in

References

[1] Yoon TP, Jacobsen EN. Privileged chiral catalysts. Science. 2003;**299**: 1691-1693

[2] Schiff H. The syntheses and characterization of Schiff base. Annals of Chemistry. 1864;**3**:343-349. DOI: 10.3390/M986

[3] Cozzi PG. Metal–Salen Schiff base complexes in catalysis: Practical aspects. Chemical Society Reviews. 2004;**33**(4): 410-421

[4] Gupta KC, Sutar AK. Catalytic activities of Schiff base transition metal complexes. Coordination Chemistry Reviews. 2008;**252**:1420-1450

[5] Arulmurugan S, Helen P, Kavitha VBR. Biological activities of Schiff base and its complexes: A review. Rasayan Journal of Chemistry. 2010;**3**: 385-410

[6] Huffaker RM. Inhomogeneous radiant heat transfer from Saturn rocket exhaust plumes. No. NASA-TM-X-53630. 1967

[7] Yamada S. Advancement in stereochemical aspects of Schiff base metal complexes. Coordination Chemistry Reviews. 1999;**190**:537-555

[8] Akhter S, Zaman HU, Mir S, Dar AM, Silaev SS. Synthesis of Schiff base metal complexes: A concise review. European Chemical Bulletin. 2017;**10**:75

[9] Akhter S, Zaman HU, Mir S, Dar AM, Silaev SS. Synthesis of Schiff base metal complexes: A concise review. European Chemical Bulletin. 2017;**6**:483

[10] Olszwske EJ, Martin DF. Interaction of amines and some nickel (II)-Schiff base compounds-II adduct formation. Journal of Inorganic Nuclear Chemistry. 1965;**27**:345-349

[11] Yamanouchi K, Yamada S. Bibliography: General literature. Biological Medicine. 1970;**34**:113-116

[12] Cox LR, Ley SV, Diederich F, Stang PJ. Templated organic synthesis. 2000

[13] Gimeno N, Vilar R. Anions as templates in coordination and supramolecular chemistry. Coordination Chemistry Reviews. 2006;**250**(23–24): 3161-3189

[14] Dalia SA, Afsan F, Hossain MS, Khan MN, Zakaria C, Zahan MKE, et al. A short review on chemistry of schiff base metal complexes and their catalytic application. International Journal of Chemical Studies. 2018;**6**:2859-2866

[15] Duatti A, Marchi A, Rossi R, Magon L, Deutsch E, Bertolasi V, et al. Reactivity of 2-(2-hydroxyphenyl) benzothiazoline with the oxotetrahalometalate (V) complexes [MOX4]-(M= technetium, rhenium; X= Cl, Br). Synthesis and characterization of new oxo-M (V) complexes containing 2-(2-hydroxyphenyl) benzothiazole. Crystal structure of tetraphenylarsonium oxotrichloro [2-(2-hydroxyphenyl) benzothiazolato] technetate (V). Inorganic Chemistry. 1988;**27**(23):4208-4213

[16] Soliman AA, Linert W. Structural features of ONS-donor salicylidene Schiff base complexes. Monatshefte für Chemie-Chemical Monthly. 2007; **138**(3):175-189

[17] Kim E, Eduardo E, Chufán KK, Karlin KD. Synthetic models for heme–copper oxidases. Chemical Reviews. 2004;**104**(2):1077-1134

[18] Meenachi S, Chitra S. A review of chemistry and biological importance of Schiff base. ChemInform. 2015;**46**:13

[19] Manjunatha KB, Supriya S, Shakeel Nawaz S, Ranjan P, Chakraborty T, Poornesh P, et al. Nonlinear optical and quantum chemical studies of palladium benzimidazole Schiff base complex. Materials Science in Semiconductor Processing. 2022;**151**:107012

[20] Pawlica D, Marszałek M, Mynarczuk G, Sieroń L, Eilmes J. New unsymmetrical Schiff base Ni (II) complexes as scaffolds for dendritic and amino acid superstructures. New Journal of Chemistry. 2004;**28**(12): 1615-1621

[21] Costes JP, Dahan F, Fernandez Fernandez MB, Fernandez Garcia MI, Garcia Deibe AM, Sanmartin J. General synthesis of 'salicylaldehyde half-unit complexes': Structural determination and use as synthon for the synthesis of dimetallic or trimetallic complexes and of 'self-assembling ligand complexes'. Inorganica Chimica Acta. 1998;**274**(1): 73-81

[22] Mukherjee P, Sengupta O, Drew MGB, Ghosh A. Anion directed template synthesis of Cu (II) complexes of a N, N, O donor mono-condensed Schiff base ligand: A molecular scaffold forming highly ordered H-bonded rectangular grids. Inorganica Chimica Acta. 2009; **362**(9):3285-3291

[23] Kaczmarek MT, Jastrząb R, Hołderna-Kędzia E, Radecka-Paryzek W. Self-assembled synthesis, characterization and antimicrobial activity of zinc (II) salicylaldimine complexes. Inorganica Chimica Acta. 2009;**362**(9):3127-3133

[24] Sumrra SH, Ibrahim M, Ambreen S, Imran M, Rehmani MDFS. Synthesis, Spectral Characterization, and Biological Evaluation of Transition Metal Complexes of Bidentate N, O Donor Schiff Bases. London: Hindawi Publishing Corporation; 2014. pp. 1-10

[25] Ambike V, Adsule S, Ahmed F, Wang Z, Afrasiabi Z, Sinn E, et al. Copper conjugates of nimesulide Schiff bases targeting VEGF, COX and Bcl-2 in pancreatic cancer cells. Journal of Inorganic Biochemistry. 2007;**101**(10): 1517-1524

[26] Patange AN, Yadav UM, Desai PA, Singare PU. Synthesis of some novel halogenated platinum (II) complexes of active Schiff's base ligand derived from 5-bromo isatin and evaluation of their antibacterial activity. World Scientific News. 2015;**10**:32-43

[27] Rosenberg B, Van Camp L, Krigas T. Inhibition of cell division in Escherichia coli by electrolysis products from a platinum electrode. Nature. 1965;**205**: 698-699

[28] Patange AN, Yadav UM, Desai PA, Singare PU. Synthesis and antimicrobial activities of novel palladium (II) complexes of active Schiff's base ligand derived from 5-bromo isatin. International Letters of Chemistry, Physics and Astronomy. 2015;**52**:22-27

[29] Pon S, Mini PS, Theodore David Manickam S, Antony R, Muthupoongodi S, Magala Sathyasheeli S. New class of Copper (II) complex derived from Isatin and Thiosemicarbazide- Synthesis. Der Pharma Chemica. 2016;**8**:67-76

[30] Fernández G, Juan M, Fred A, et al. The structures and cyclic voltammetry of three copper (II) complexes derived from bulky ortho-hydroxy Schiff bases. Journal of Molecular Structure. 2002; **612**(1):69-79

[31] Baleizao C, Garcia H. Chiral salen complexes: An overview to recoverable and reusable homogeneous and heterogeneous catalysts. Chemical Reviews. 2006;**106**(9):3987-4043

[32] Gogoi A, Sarmah G, Dewan A, Bora U. Unique copper–salen complex: An efficient catalyst for N-arylations of anilines and imidazoles at room temperature. Tetrahedron Letters. 2014; **55**(1):31-35

[33] Jacobsen EN, Kakiuchi F, Konsler RG, Larrow JF, Tokunaga M. Enantioselective catalytic ring opening of epoxides with carboxylic acids. Tetrahedron Letters. 1997;**38**(5):773-776

[34] Gaponik PN, Voitekhovich SV, Ivashkevich OA. Metal derivities of tetrazoles. Russian Chemical Reviews. 2006;**75**(6):507-539

[35] Rad MNS. An efficient protocol for facile synthesis of new 5-substituted-1H-tetrazole derivatives using copper-doped silica cuprous sulfate (CDSCS) as heterogeneous nano-catalyst. Journal of the Brazilian Chemical Society. 2017;**28**: 11-20

[36] Xie A, Cao M, Liu Y, Feng L, Xinyu H, Dong W. The synthesis of tetrazoles in nanometer aqueous micelles at room temperature. European Journal of Organic Chemistry. 2014;**2014**(2): 436-441

[37] Lauder K, Toscani A, Scalacci N, Castagnolo D. Synthesis and reactivity of propargylamines in organic chemistry. Chemical Reviews. 2017;**117**(24): 14091-14200

[38] Li P, Regati S, Huang H-C, et al. A sulfonate-based Cu (I) metal-organic framework as a highly efficient and reusable catalyst for the synthesis of propargylamines under solvent-free conditions. Chinese Chemical Letters. 2015;**26**(1):6-10

[39] Agrahari B, Layek S, Ganguly R. Synthesis and crystal structures of salen-type Cu (ii) and Ni (ii) Schiff base complexes: Application in [3+ 2]-cycloaddition and a 3-coupling reactions. New Journal of Chemistry. 2018;**42**(16):13754-13762

[40] Al-Amin M et al. Selectivity, compatibility, downstream functionalization, and silver effect in the gold and palladium dual-catalytic synthesis of lactones. Organometallics. 2014;**33**(19):5448-5456

[41] Li GY. The first phosphine oxide ligand precursors for transition metal catalyzed cross-coupling reactions: C– C, C– N, and C– S bond formation on Unactivated aryl chlorides. Angewandte Chemie International Edition. 2001;**40**(8):1513-1516

[42] Suganthy PK, Prabhu RN, Sridevi VS. Synthesis, structure and catalytic applications of octahedral nickel (II) benzoylhydrazone complex: Suzuki–Miyaura cross-coupling reaction of aryl bromides with arylboronic acid. Inorganica Chimica Acta. 2016;**449**:127-132

[43] Bazzan I, Volpe A, Dolbecq A, Natali M, Sartorel A, Mialane P, et al. Cobalt based water oxidation catalysis with photogenerated Ru (bpy) 33+: Different kinetics and competent species starting from a molecular polyoxometalate and metal oxide nanoparticles capped with a bisphosphonate alendronate pendant. Catalysis Today. 2017;**290**:39-50

[44] Ansari RM, Mahesh LK, Bhat BR. Cobalt Schiff base complexes: Synthesis characterization and catalytic application in Suzuki–Miyaura reaction. 2019;**2019**:556-563

[45] Abdel-Fatah SM, Díaz-Sánchez M, Díaz-García D, Prashar S, Abdel-Rahman LH, Gómez-Ruiz S. Nanostructured metal oxides prepared from schiff base metal complexes: Study of the catalytic activity in selective oxidation and C–C coupling reactions. Journal of Inorganic and Organometallic Polymers and Materials. 2020;**30**(4):1293-1305

[46] Buldurun K, Özdemir M. Ruthenium (II) complexes with pyridine-based Schiff base ligands: Synthesis, structural characterization and catalytic hydrogenation of ketones. Journal of Molecular Structure. 2012;**2020**:127266

[47] Fuchs M, Schmitz S, Schäfer PM, Secker T, Metz A, Ksiazkiewicz AN, et al. Mononuclear zinc (II) Schiff base complexes as catalysts for the ring-opening polymerization of lactide. European Polymer Journal. 2020;**122**:109302

[48] Niakan M, Karimi S, Masteri-Farahani M, Shekaari H. An efficient, cost-effective, and magnetically recoverable copper catalyst for O-arylation of phenols with aryl halides in choline chloride-based deep eutectic solvents. Colloids and Surfaces A: Physicochemical and Engineering Aspects. 2021;**620**:126603

[49] Bunno Y, Tsukimawashi Y, Kojima M, Yoshino T, Matsunaga S. Metal-containing Schiff Base/Sulfoxide Ligands for Pd (II)-Catalyzed Asymmetric Allylic C–H Aminations. ACS Catalysis. 2021;**11**(5):2663-2668

[50] Kumar A, Verma S. Synthesis and characterization of a recyclable graphene oxide-surface-engineered copper (II) Schiff base complex: Catalytic application in synthesis of 1, 2, 3-triazoles and 2H-indazoles. Journal of Environmental Chemical Engineering. 2021;**9**(4):105791

[51] Ramakrishna D, Bhat BR. Cobalt complexes in [EMIM] Cl–A catalyst for oxidation of alcohols to carbonyls. Inorganic Chemistry Communications. 2010;**13**(1):195

[52] Ramakrishna D, Bhat BR. Catalytic oxidation of alcohols by nickel (II) Schiff base complexes containing triphenylphosphine in ionic liquid: An attempt towards green oxidation process. Catalysis Communications. 2010;**11**(5):498

[53] Dileep R, Ramachandra Bhat B. Palladium-Schiff base-triphenylphosphine catalyzed oxidation of alcohols. Applied Organometallic Chemistry. 2010;**9**(24):663-666

[54] Ramakrishna D. Green conversion of alcohols to carbonyls catalyzed by novel ruthenium-Schiff base-triphenylphosphine complexes. Inorganic Chemistry Communications. 2011;**14**(1):155

[55] Ramakrishna D, Bhat BR. A catalytic process for the selective oxidation of alcohols by copper (II) complexes. Inorganic Chemistry Communications. 2011;**14**(5):690

[56] Dileep R. Palladium complex in a room temperature ionic liquid: A convenient recyclable reagent for catalytic oxidation. Green Chemistry Letters and Reviews. 2014;**7**(1):32

[57] Dileep R, Rudresha BJ. An ionic liquid immobilized copper complex for catalytic epoxidation. RSC Advances. 2015;**5**(81):65870-65873

[58] Ramakrishna D. An ionic liquid immobilized palladium complex for an effective heck and Suzuki coupling

reactions. Chemistry Africa. 2019;**2**(1): 21-28

[59] Ramakrishna D. Ionic liquid immobilized Cu (I)–hydrazone–triphenylphosphine complex: An easily recyclable catalytic system for Suzuki and Heck cross couplings. Canadian Journal of Chemistry. 2021;**99**(6): 497-501

[60] Ramakrishna D, Saravu S, Rajendrachari S. A palladium complex dispersed in ionic liquid as an efficient catalytic combination for the synthesis of Benzazoles. Topics in Catalysis. 2022; **2022**:1-11